John Alva Myers

Experiments upon wheat

John Alva Myers

Experiments upon wheat

ISBN/EAN: 9783743414822

Manufactured in Europe, USA, Canada, Australia, Japa

Cover: Foto ©berggeist007 / pixelio.de

Manufactured and distributed by brebook publishing software
(www.brebook.com)

John Alva Myers

Experiments upon wheat

Bulletin No. 7

WEST VIRGINIA

Agricultural Experiment Station.

— § —

I. EXPERIMENTS UPON WHEAT.
II. EXPERIMENTS UPON FRUIT TREES.
III. EXPERIMENTS UPON GARDEN SEEDS, &C.
IV. EXPERIMENTS UPON GRASSES AND FORAGE CROPS.
V. EXPERIMENTS UPON MISCELLANEOUS SUBJECTS.

— § —

By JOHN A. MYERS,
Director.

— § —

CHARLESTON:
MOSES W. DONNALLY, PUBLIC PRINTER.
1890.

BOARD OF REGENTS OF THE WEST VIRGINIA UNIVERSITY.

District.	Name of Regent.	P. O. Address.
1.	J. B. SOMMERVILLE,	Wheeling.
2.	CLARENCE L. SMITH,	Fairmont.
3.	PEREGRIN HAYES,	Glenville.
4.	D. D. JOHNSON,	Long Reach.
5.	JOHN G. SCHILLING,	Spencer.
6.	EDWARD A. BENNETT,	Huntington.
7.	WIRT A. FRENCH,	Princeton.
8.	M. J. KESTER,	Union.
9.	D. C. GALLAHER,	Charleston.
10.	THOMAS J. FARNSWORTH,	Buckhannon.
11.	JOSEPH MORELAND,	Morgantown.
12.	JOHN A. ROBINSON,	Patterson's Depot.
13.	DR. W. W. BROWN,	Kabletown.

MEMBERS OF THE STATION COMMITTEE.

JOHN A. ROBINSON,   PEREGRIN HAYS,
JOSEPH MORELAND,   THOMAS J. FARNSWORTH,
DR. W. W. BROWN,   JOHN A. MYERS.

PRESIDENT OF THE UNIVERSITY.   TREASURER.

E. M. TURNER, LL. D.,   JOHN I. HARVEY.

STATION STAFF.

JOHN A. MYERS, PH. D., - - - - - Director.
A. C. MAGRUDER, B. S., - - - - Creamery-man.
H. R. BALDWIN, Jr., - - - - - - Chemist.
DR. CHAS. F. MILLSPAUGH, - Botanist and Microscopist.
SUSIE V. MAYERS, - - Stenographer and Book-keeper.

The work reported in this Bulletin is that carried out by the direction of our Board of Regents under the following resolution, which was passed at the meeting of the Board June 12, 1887.

Order of the Board.

"Ordered, that the Director of the Agricultural Experiment Station, upon the application of the Regent of any Senatorial District in the State, is authorized and directed to expend in making experiments in Agriculture, Horticulture and Veterinary within such Senatorial District the sum of $600; which experiments are to be made under the directions of said Director acting in conjunction with the Regent of the District in which such experiments are made. Full reports thereof shall be made by said Director to be printed and circulated throughout the State. And upon application of any Regent said Director shall furnish to him as many copies of said reports, with all papers and exhibits therewith, as he shall require, properly folded and stamped ready for mailing, or as otherwise required, in order that such Regent may send them post-paid to any address he may desire.

And if such reports shall be furnished to any regent without being stamped, he is authorized and requested to circulate them by mail to citizens of West Virginia, and will be paid such sums as he shall expend for postage, etc., out of the funds of the Station."

In accordance with this order, the following circular letter was sent to each member of the Board of Regents:

MORGANTOWN, W. VA., Sept. 6th, 1888.

Dear Sir:

In carrying out the policy of the Board of Regents in regard to having experiments conducted in each of the Senatorial Districts, it appears to be desirable to devote a share of our attention to the wheat crop, forming as it does an important source of revenue to the farmers in many parts of the state. It appears to be advisable to try what can be accomplished in the way of trial of new or untried varieties. Should it meet with your approval, therefore, I will cause several varieties of seed wheat to be sent into your District freight prepaid, to such freight stations as you may designate, and to such persons as you may name, not exceeding fifty. The seed will be delivered in 1¼ bushel sacks, which is considered the proper amount of seed to sow to the acre. This will make all of the experiments of sufficient size to make it an object for the farmers to undertake the work. They are to have the crop as their reward for making the trial and reporting upon the folllowing' points, i. e.: The time and manner of sowing, the character of the soil and its manner of preparation with previous crops grown upon it; the state of the weather at the time and following the planting, the ravages of insects and diseases, the ability of the wheat to resist the winter, and drought, should we have any, the peculiarities of growth and the quantity harvested.

The selection of the experimenters is left wholly to the good judgment of the local regent, and the choice of plats must also be settled by him and the farmer. It is not within the power of the Director to visit all sections of the State in time to assist in any of the work this Fall. It should be properly impressed upon the persons who engage in the work that it is not intended that the seed is given to them free of all obligations.

If they fail to report upon the work, it is simply so much of the funds of the Station diverted from their proper use, as they are intended to give the farmers the benefit of the experience derived from their proper expenditure. It is hoped, therefore, that care will be exercised to see that those who use the seed will give the Station the benefit of their experience, for the good of the public, as the results will be published.

Please inform me what varieties of wheat are now generally used in your district, so that new varieties may be introduced, as little would be gained by sending old varieties. It is now close to seeding time, and I trust that the gentlemen who receive this will give it immediate attention, as otherwise it will be too late. If desired, I will send the wheat directly to the Regent himself, and he may distribute it as he deems for the best.

Respectfully,

JOHN A. MYERS,
Director.

Letter sent to each Farmer to whom Seed was sent.

WEST VIRGINIA EXPERIMENT STATION,
MORGANTOWN, W. VA., ..., 188....

DEAR SIR:

By the direction of ....
the Regent of the University from your Senatorial District, I send
you...
....
....
....
....
....
....
.... by ....
to ....where you will probably
find it in due time. Should there be any transportation charges
upon it when it arrives at that point, please pay the same and send
me the freight bill, upon receipt of which I will refund the charges
to you, as the circumstances may be such as to prevent pre-payment
of the same at the time of shipment.

You will find sufficient in each package for one acre of ground.
This is not intended as a present, but you will confer a favor upon
the community, and upon the State by making such simple notes
and observations for publication as may be called for by the Director of the Experiment Station, together with such matters of interest as you may see fit to add thereto. It is intended for the benefit
of farmers throughout the State, and should not be held back.

The Experiment Station is put to a very considerable expense in
carrying out this line of experimental work, and its success must
depend upon whether the farmers whom it thus attempts to benefit,
will meet it half way and freely contribute their experience for the
benefit of their fellow-craftsmen.

It is our opinon that if Agriculture is to be elevated and made
more profitable it must be brought about by combining the experience of intelligent farmers like yourself, with the careful observations of scientific men in such a manner as to form a more exact
foundation upon which to build.

You are therefore invited to help us by contributing such information as may be available, and have been named to me as a suitable person to rely upon.

Yours truly,

JOHN A. MYERS,
Director,

This was also accompanied by the circular giving directions for
conducting experiments, as follows:

Direction for Testing the Adaptability of Grains, Grasses, or Fertilizers to the Soils of West Virginia.

Persons to whom seeds of any kind or fertilizers are sent will find them contained in packages of sufficient quantity to be applied to an acre of ground. They are prepared especially with reference to application to an acre of ground.

The farmer will prepare the ground in the usual manner for such crops where seed is to be sown, and plant it in the manner usually adopted in that section of the country. In the case of fertilizers he can apply them either by broad-casting them, by drilling them, or by applying them near (*not against*) the plant. In the latter case the farmers frequently make the mistake of placing concentrated fertilizers too close to the plant. The fertilizers sent out should be applied to the surface of the ground, and at most not more than harrowed, or lightly drilled in. They are intended to be *quick acting*, and as the crops to which they are applied generally have the feeding power of their roots distributed near the surface of the soil, it is advisable to allow the fertilizer to be soaked down by the action of the rain, rather than plow it in so deep that the plant will never get the benefit of it.

In making experiments we must have exact quantities for the sake of comparison. By using any of the following dimensions an acre of ground may be readily staked off. All that is required is simply a rope or tape line of any convenient length, say 33 feet or 2 rods long. With this measure off any of the longer lines, say 40 rods. Then from the ends of this line run out the shorter line, say 4 rods, and set in stakes. Observe care to have opposite sides parallel.

10 rods	×	16 rods	= 1 acre.
8 "	×	20 "	= 1 "
5 "	×	32 "	= 1 "
4 "	×	40 "	= 1 "
5 yards	×	968 yds.	= 1 "
10 "	×	484 "	= 1 "
20 "	×	242 "	= 1 "
40 "	×	121 "	= 1 "
80 "	×	$60\frac{1}{2}$ "	= 1 "
70 "	×	$69\frac{1}{2}$ "	= 1 "
220 feet	×	198 feet	= 1 "
440 "	×	99 "	= 1 "
110 "	×	369 "	= 1 "
60 "	×	726 "	= 1 "
120 "	×	363 "	= 1 "
240 "	×	$181\frac{1}{2}$ "	= 1 "

The farmer should make note of and report to the Director the *manner* in which the soil *is prepared* for the crop, the *previous condition* of the soil with reference to the crops grown upon it, the *condition of the soil at the time of planting,* whether in good *physical condition* or not, the *condition* of the *weather succeeding* the *time of planting*, and *success* in *getting* a *stand*, the *onslaught* of *insects* or *diseases upon the crop*, the *ability* of the *crop* to *withstand* the climate, that is *the percentage* that is able to withstand *drought* or *excessive freezing*, the *peculiarities of the growth of the crop*, *time of ripening*, and the *yield per acre secured*.

These specimens are sent out with the expectation that all the farmers of the State will get the benefit of the experience derived from them, and we must rely upon the intelligence and good judgment of the farmers doing the work for any benefit that the farmers of the State are to derive from them. All the work required in this case is of the simplest possible character. The land should be measured before it is planted, and the yield weighed or measured after it is harvested. As reward for his work, the farmer's report will be published, and the harvest secured will belong to him.

The Station sends out these samples in sufficient bulk to make it an object for the farmer to undertake the work. It is the desire of the authorities of the Station that the farmers mutually support one another. It is hoped that the farmers will freely consult the authorities of the Station and communicate with them concerning all the difficulties as well as the successes that they meet with in their business. Where they have success of any kind it is desirable that others should have the benefit of their experience. Where they meet with failure and disappointment it is desired that we should know it, in order that we may inquire into the causes, and if possible suggest some means of remedying it.

We hope that all will avail themselves of such facilities as the Station may be able to offer, and its Director together with his staff of assistants will at all times stand ready to serve the interests o the farmers.

JOHN A. MYERS,
Director,

I. WHEAT.

The following are the varieties of wheat sent out to be tested:

Michigan Bronze.	Rice.
Solid Gold.	White Booten.
Velvet Chaff.	Fultz.
Golden Amber.	Hybrid Mediterranean.
Nail Club.	Red Russian.
Lancaster.	Deitz Longberry.
German Amber.	Reliable.
Purple Straw.	Tuscan Island.
Findlay.	Valley.
Raub's Black Prolific.	Good.

FIRST DISTRICT.

Variety of wheat tested, "Solid Gold."
Quality of seed, good.

Sowed 1½ bushels per acre, October 27, 1888, on limestone soil, prepared by plowing and harrowing. Condition of soil at time of sowing was very wet; too wet for using a drill. The land in 1887 was in corn. Fair yield. The soil is considered well adapted for wheat. The weather following sowing was very wet, and the stand of wheat secured in the fall was very poor. It was sown so late that the wheat did not get a good start in the fall. Ripened June 20th. Yielded 16 bushels per acre; very nice wheat. Not such a good grain as the " Velvet Chaff," but a very good wheat. Like it quite well, and think it well adapted to this country.

WILLIAM FLAHERTY,
West Liberty, Ohio county, W. Va.

FIRST DISTRICT.

Variety of wheat tested, "Michigan Bronze."
Quality of seed, fine.

Sowed, 1¼ bushels per acre, October 27, 1888, on limestone soil, prepared by plowing and harrowing. It was very wet at time of sowing; too wet for the drill. The land in 1887 was in corn; fair yield. The soil is considered well adapted to wheat. The weather following sowing was very wet, and stand of wheat secured in the fall was very poor; some of the wheat not coming up until spring. It withstood the winter very well. Ripened July 4th; yield, 20 bushels per acre. The wheat withstands drought and cold, but not wet weather. It grew in shock, while under the same conditions the other did not. Very small grain; do not like it very well.

WILLIAM FLAHERTY,
West Liberty, Ohio county, W. Va.

FIRST DISTRICT.

Variety of wheat sown, "Velvet Chaff."
Quality of seed, excellent.

Sowed 1½ bushels per acre, October 27, 1888, on limestone soil, prepared by plowing and harrowing. Condition of soil at time of sowing was very wet; too wet to use the drill. Land in corn in 1887; fair yield. It is considered well adapted to wheat. Weather following the sowing was remarkably wet and cold. The stand of wheat secured in the fall was not at all good. It is considered by us the best wheat there is to withstand cold. Ripened about June 20th; yield, 16 bushels per acre. This was the earliest wheat I had. The small yield, I think, was due to the late sowing; in consequence of which it got a poor start in the fall, some not coming up until in the spring. I think it the wheat best adapted to this section of the country.

WILLIAM FLAHERTY,
West Liberty, Ohio county, W. Va.

FIRST DISTRICT.

Variety of wheat tested, " Golden Amber."
Quality of seed, good.
Sowed 1¼ bushels per acre.
Soil prepared same as other varieties ; all conditions the same.
Stand of wheat secured in the fall the same.
Peculiarities of crop : Large, plump, full heads.   Straw almost like hazel brush ; so stiff you could hardly bind it.   Harvested same day as other two ; yield was 19½ bushels, measured at machine. It grew very tall.   A very hard, plump grain.   Number of dozen 37, average bind.

A. R. JACOB,
Clinton, W. Va.

FIRST DISTRICT.

Variety of Wheat, "Michigan Bronze,
Quality of seed very good.   Sowed 1 bushel per acre October 10th., 1888, on limestone, clay soil prepared by plowing, harrowing and pulverizing.   Good physical condition of soil at time of sowing.

The land in 1887 was in meadow.   Yield 2 to 2½ tons of Timothy Hay per acre.   In 1886 in Timothy and clover.   In 1885 in Timothy and clover.   Soil considered well adapted for wheat.   The weather following sowing was wet, cold and unfavorable.   The stand of wheat secured in the fall was poor, being sown so late, Not so good to withstand frost and drought as "Velvet Chaff." It is thinner on the ground.   Rather stiff straw with smaller head, Harvested July 8th,   Latest of the three kinds,   Yield 18 bushels per acre,   Measured at machine,   A good hard wheat,   Number of dozens 30 average bind.   Yielded very well to bulk of straw.

A. R. JACOBS,
Clinton, Ohio County, W. Va.

VELVET CHAFF.

Drilling in wheat, Sowed, the fertilizers received broad-cast, then gave it a stroke with my slant-tooth harrow, slightly covering the fertilizer.   I had to sow in corn ground, as I had given up all hope of receiving seed in time to sow last fall ; when it came, so late I thought 1 would take the risk.   Of the three different kinds of fertilizer used viz : Royal Bone Phosphate, Orchilla Guano and

Schenk's Fertilizer, my observation was that Schenks' fertilizer gave the best result. We have sown 6 bushels of the different kinds of seeds this fall, and hope to be able to make a more satisfactory report of our test after next harvest.

A. R. JACOBS.
Clinton, W. Va.

FIRST DISTRICT.

Variety of wheat tested, " Velvet Chaff."
Quality of seed, very good.

Sowed 1¼ bushels per acre, October 15, 1888, on sandy loam soil, prepared by first plowing deep, then thoroughly harrowed and thoroughly compacted with a heavy drag. Condition of soil at time of sowing, very good, excepting a little wet. The same land produced 250 bushels potatoes per acre in 1887; in 1886 and 1885 in pasture ; the soil is considere l very good for wheat. It was very wet following time of sowing, and scarcely any wheat got through the ground until spring. About one-third of the plat was drowned out by water standing on it. Cannot tell as to the ability of the wheat to withstand frost and drought, as no wheat was frozen or dried out here this season. It is a remarkable wheat to stool out. Wheat was cut with binder July 8th. The yield on ⅔ acre was 14 bushels. The grain is very plump and hard. Should make excellent flour, but think the straw too soft.

L. C. APPLEGATE.
Wellsburg, Brooke county, W. Va.

FIRST DISTRCT.

Varieties tested : Golden Amber, Michigan Bronze and Velvet Chaff.

Sowed 1¼ bushels per acre October 24th., 1888, on rather saudy soil prepared as usual. It rained and kept the ground too wet, so that it did not look well in the fall. The soil at time of sowing was tolerably dry, but very wet afterwards. The soil is considered well adapted to growing wheat. Yield was 10 bushels per acre for Golden Amber. For Michigan Bronze 8 bushels per acre. For Velvet Chaff 7 bushels per acre. The latter did no good. I think the Golden Amber will do here, and the Michigan Bronze is good, but the Velvet Chaff will not do here. I will try the two kinds again,

JESSE HUKILL,
Wellsburg, Brooke County, W. Va

FIRST DISTRICT.

Variety of wheat tested "Velvet Chaff."
Quality of seed good.

Sowed 1¼ bushels per acre October 10th, 1888, on limestone clay soil, prepared by plowing, harrowing and pulverizing as fine as we could ; being corn stubble. Soil at time of sowing in good physical condition. Crop grown in 1887 was timothy hay. In 1886 timothy and clover. In 1885 timothy and clover. Yield 2 to 2½ tons per acre.

The adaptation of the soil to wheat is considered good, being a rich limestone clay. The weather following the sowing was wet and cold, and very unfavorable. The stand of wheat secured in the fall was very poor; being sown so late. The wheat lodged in patches, and English sparrows destroyed some heads, being sown near buildings. The ability of the crop to withstand frost and drought very good. It is rather a soft straw with large, well filled head. Harvested July 8th. Very ripe. Yield per acre 24½ bushels measured at machine.

Other observations: You will remember that the seed was caught in the great railroad blockade, so that I did not receive it until the morning of the 10th of October, and sowed in the afternoon. This being so late in the season, I can not consider it a fair test. Considering everything, I think the yield good.

Variety of wheat "Golden Amber."

Quality very fair, with some imperfect grains. Number of dozen 45 average bind. I measured off an acre by your instructions in table.

Sowed 1¼ bushels November 1st, 1888, on sandy land with clay subsoil prepared by plowing, harrowing, dragging harrowing, dragging and harrowing. The soil was in extra fine condition, but very wet at the time of sowing. The same land in 1887 produced 150 bushels potatoes per acre. In 1886, 200 bushels potatoes per acre. In 1885 in pasture. Soil is considered good. The wheat was drilled, and the weather following was cold and wet. The stand of wheat secured in the fall was very poor, owing to the late sowing and wet weather. No accidents or diseases noticed. Ripens from July 5th to 10th. Did not separate from other varieties of wheat in threshing. Early sown on good ground would likely make a favorable impression. The late sowing and wet season, were so much against the crop doing well, that opinions expressed from my past experience would be guesses.

S. C. GIST.

Wellsburg, Brooke county, W. Va.

RECAPITULATION FOR FIRST DISTRICT.

"VELVET CHAFF."

Excellent wheat. Well adapted to the this section.
Considered an excellent wheat.
Not adapted to this section.
Think it well adapted to this section.

GOLDEN AMBER.

Considered good.
Uncertain.
Considered adapted to this section.

"SOLID GOLD."

Not so good as "Velvet Chaff," but a very excellent wheat.

"MICHIGAN BRONZE."

A good hard wheat.
Good.
Grew in shock and under similar conditions other wheat did not.
Not liked.

90 tests were sent to th's D strict, from which 12 reports were received. Of this number the reports were as follows:

"Velvet Chaff."   3 reports were favorable.
1 reports unfavorable.
"Golden Amber,"   2 were favorable.
1 report were unfavorable.
"Solid Gold,"   1 report favorable.
0 reports unfavorable.
"Michigan Bronze," 3 reports favorable
1 report unfavorable.

It will be remembered that the blockade in railroad freights prevented, perhaps, one-third of the parties to whom the seed whea was sent from securing it in time for seeding. Other parties, we presume, we shall never hear from.

SECOND DISTRICT.

Variety of wheat tested "Michigan Bronze."
Quality of seed was very good.
Sowed 1¼ bushels per acre November 1st, 1888, on sandy, clay soil, prepared by plowing and harrowing. The soil at time of sowing was wet and the weather cold. The land in 1887 was in oats. Yield was about 20 bushels per acre. In 1886 it was in corn. Soil is considered low. Weather following seeding was cold and wet. Stand of wheat secured in the fall was very poor owing to the condition and time of sowing. It did not start growing until some time in the winter; the ground not settling until freezing. Think this variety will stand frost well. It is a short, thick straw. Short, thick heads, red chaff and beards, and large red grain. Time of ripening July 2d. Yield 12 bushels per acre measured at machine. It is a good wheat for low, wet lands. One small piece of this plat was a black loam and shaded some. I sowed "Fultz" and "Pool" and "Michigan Bronze" side by side in this piece the same day. "Fultz" and "Pool" did not make anything. "Michigan Bronze"

made a good crop. I think it would make wheat to sow in the woods (fresh land).

JESSE L. BONAR,
Glen Easton, Marshall county, W. Va·

SECOND DISTRICT.

Variety of wheat tested "Solid Gold."
Quality of seed was good.

Sowed 2 acres 1¼ bushels per acre, October 11th., 1888, on clay soil, prepared by plowing, harrowing aad dragging to make a fine seed bed. The soil at time of sowing was very wet, and rained all the time of drilling in the wheat. The same land produced in 1887 eight bushels ot wheat per acre, In 1885 and 1886. it was in grass. The soil is not considered very good for wheat, and the weather following sowing was very wet. The stand of wheat secured in the fall was very good; being sown late, and in the mud. Noticed no accidents, diseases nor other damage to the crop. This variety stood the weather of June very well, and also stood the frost ot winter well. It is a hearty looking plant during winter, with nice, golden bright straw. It ripened 9th. of July. The yield was 18½ bushels per acre from machine. Is a long straw. A good variety for thin land. Would not advise sowing on rich, black soil, or on low lands, Making long straw, it would tangle and not fill. Can not be beat for high thin lands. It is the best variety we have in our section for the high, thin lands. Has a large red grain with long heads and straw.

JESSE L. BONAR,
Glen Easton, Marshall County, W. Va.

SECOND DISTRICT.

Name of variety "Nail Club."
Quality of seed not very good,

Sowed 1¼ bushels per acre October 11th., 1888, on clay soil containing a little soapstone, prepared by plowing, harrowing and dragging to make a fine seed bed. The soil at time of sowing was wet. The weather in this section being wet it was not a good seed time. In 1887 the same land produced about 8 bushels wheat per acre. In 1886 it was in grass. The weather following sowing was very wet, and the stand of wheat secured in the fall, was good considering the wet time of sowing. The wet weather the last ot June 1889, damaged the heads, and being compact and holding the water seemed to kill the chaff. Believe the variety will stand frost well. It is a fine plant during winter, growing a very short stiff straw with short close heads, ripening 9th., July. It yielded 17½ bushels per acre from machine. About ½ acre of the plant was very poor and light and the wheat did not do anything. I think I would have had as much wheat if this place had not been sown. I think the wheat is a good variety ; being a small, plump grain. By compar-

ing the wheat received last fall with what I grew from this seed, I
see it improving, or what I have grown is a better quality. Making
it short, the Nail Club "will yield more," everything being the
same, than any other wheat I have ever grown.

JESSE L. BONAR,
Glen Easton, Marshall County, W. Va.

FAIRMONT, W. VA., Nov..19, 1889.

MY DEAR SIR :

The eighteen bags sent me last fall were distributed as follows :
In Paw Paw district of this county one bag to J. A Floyd, one
to John Straight, one to Z. G. Morgan and two to O. P. Floyd.
In Mannington district, one bag to Jacob Myers, one to J. J. Mur-
ray, one bag to Fleming Hamilton and one bag to A. Z. Riggs.
In Fairmont district, one bag to Charles Conaway, one bag to
Thomas Conaway, one bag to B. D. Fleming, one bag to John B.
Gray, one bag to M. L. Fleming, one bag to E. Hamilton and one
bag to Eugenius Wilson. Lincoln district, one bag to S. E. Fleming, and one bag to T. R. Hall, in Winfield district.

The wheat was delivered to all except Messrs. B. D. and M. L.
Fleming, who failed to get the bags left for them from the depot,
which fact did not come to my knowledge until in the spring
when I had the two bags taken home and this fall had them carefully seeded by a neighbor. I would rather you would turn the
price of those two bags into the Station funds, and draw on me
for the amount. I have been informed that the wheat sent to
Mannington district, was received too late to be seeded and was
not, in fact, sown until this fall. This I only have from hearsay,
as I have failed to hear personally from any of the parties. I enclose you the only report I have received; namely, from Captain
Gray. I hope to hear from some others within a few days. I expected a very careful report from O. P. Floyd, who was one of our
best farmers and a very painstaking man, but he died about the
time the wheat was harvested. I regret very much the delay in
this matter, but I could not help it.

Very truly yours,
C. L. SMITH.

SECOND DISTRICT.

Variety of wheat tested, "Lancaster."
Quality of seed was good.
Sowed 1 bushel per acre November 8th., on tough, white oak
soil, prepared in the usual way. Condition of soil at time of sowing was very wet. Corn was growing on the land in 1887. Yield
50 bushels per acre. In 1886 in grass, in 1885, grass. Soil is considered medium. Wheather following sowing was very wet.
Stand of wheat secured in the fall was thin. Crop stood frost
very well, best of any I had. Crop grew tall. Ripened July 4th.
Yield 19 bushels per acre.
I think it a good wheat.

JNO. B. GRAY,
Fairmont, W. Va.

RECEPITULATION FOR THE SECOND DISTRICT.

"NAIL CLUB."

Considered the best wheat ever grown.

"SOLID GOLD."

Considered excellent for high, thin land.

"MICHIGAN BRONZE."

Better than either "Fultz" or "Pool."

"LANCASTER."

Considered a good wheat.

24 tests were sent to this district, of which 4 were reported upon as follows:

"Nail Club."	1 report favorable.
	0 reports unfavorable.
"Solid Gold"	1 report favorable.
	0 reports unfavorable.
"Michigan Bonze"	1 report favorable.
	0 reports unfavorable.
"Lancaster."	1 report favorable.
	0 reports unfavorable.

THIRD DISTRICT.

Variety of wheat tested "German Amber."
Quality of seed received was very good.
Sowed 1¼ bushels November 1, 1888, on yellow clay soil, prepared by plowing in the wheat with a shovel plow. The ground was too wet to use another plow. The condition of soil at time of sowing was not at all good, being too wet. The land in 1887 was in corn. Yield 50 bushels per acre. In grass the year before. The soil is considered a very good clay for wheat. Weather following sowing was rather rainy. No hard freezing during the winter. Stand secured in the fall was very good, but scarcely made a show before March. So far as I was able to observe there was no fly, rust, or anything, to interfere with or attack the wheat. It stood the winter well, and I will sow it this fall. It grew vigorously the whole spring. Ripened July 4th. Yield nineteen bushels. It is a splendid wheat, in my opinion, for this locality. A number of persons inquired as to the name of the wheat, and admired it from first to last. Think it an excellent, hardy wheat. Yield big, but the fall was so wet that I was delayed in sowing more than a month. It ought to be sown in October or September. I intend to sow 10 bushels in bottom land the present month.

D. BASSEL.

Lost Creek, Harrison county, W. Vn.

Variety of wheat tested "Purple Straw."
Quality of seed good.
Sowed 1¼ bushels per acre October 31st, 1888, on red clay soil prepared by plowing the wheat in with a shovel plow. Weather

wet at time of sowing. The land in corn in 1887. Yield 50 bushels per acre. In 1886 and 1885 in grass. Soil is considered a good, stiff, red clay, adapted to wheat. Weather following sowing was rather wet, and stand of wheat secured was poor, as it could not be seen until March. In April, the fly was in it to some extent. It was not a fair test; it being sown so late. Noticed nothing peculiar about the growth of crop. It ripened from 4th to 7th of July. Yield 14 bushels per acre. It is the old blue stem wheat that was raised in this country thirty years ago, and was a fine variety of wheat, but it run out and failed utterly. Forty years ago, it was considered the best wheat grown in this country. I believe it will be a success at this time sown on new upland. The truth is, that the fall was so wet, that it was not sown at the proper time and no fair test was made.

D. BASSEL.
Lost Creek, Harrison county, W. Va.

THIRD DISTRICT.

Variety of seed received "Purple Straw."
Quality of seed was not good. Mixed with Cockle and Rye.
Sowed 1¼ bushels per acre on light clay soil October 31., 1888. Soil was prepared by plowing well. Ground was in good condition; though very wet at time of sowing. It was sod grown, and had been for 12 years. The soil is considered not adapted to wheat. The weather following time of sowing was rainy. The stand of wheat secured in the falir was fair. Froze out some in the winter. Its ability to withsand frost is not good. No special peculiarities noticed. Time of ripening July 1st. Yield 5 bushels per acre. I did not receive the seed in time to give it a fair test. Did not receive it until the 1st., of October. I believe the wheat is good, if given a fair chance.

W. PRUNTY,
Goff's, Ritchie County, W. Va.

RECAPITULATION FOR THE THIRD DISTRICT.

"GERMAN AMBER."

Considered an excellent wheat.

"PURPLE STRAW."

Believed to be a good wheat.

72 tests were sent to this district, of which
3 were reported upon as follows :

"German Amber." 1 report favorable.
0 reports unfavorable.

"Purple Straw" 2 reports favorable.
0 reports unfavorable.

169

Variety of wheat tested, "Golden Amber."
Quality of seed received, fair.
Sowed 1¼ bu. per acre Oct. 11th 1888. The soil is river bottom,
clay loam, prepared by two-horse plow and harrowed. The condition of the soil at time of sowing was not good. Plowed well,
but was wet when sown. In 1887, the land was in clover. In
1886 in clover. In 1885 in wheat. Yield 15 bu. per acre. Soil
is considered good for wheat. Weather following sowing was cold
and wet. The stand secured in the fall was tolerably good. The
crop withstands frost and drought well. It grows very tall, with
long, white heads. Ripened July 1st. Yield 23½ bu. per acre.
It is very tall wheat, with strong straw. Stands wind well. I
consider it a very good wheat.

PHILIP W. WELLS,
Long Reach, Tyler Co., W. Va.

FOURTH DISTRICT.

Variety of wheat tested, "Velvet Chaff."
Quality of wheat, fair.
Sowed 1½ bu. per acre October 11th, 1888. Soil, river bottom,
clay loam, prepared by two-horse plow and harrow. It was a
good chance for wheat. The soil was good, but at time of sowing
it was wet. In 1887, it was in clover. In 1885, in wheat. Yield
15 bu. per acre. Soil is considered good for wheat. It was
sown broadcast. Stand secured in the fall was tolerably good.
Wind blew one-third of it down two weeks before harvest.
Stands frost and drought well. Peculiarities of growth—Dark
green blades, with velvet chaff. Very tall. Ripens july 1st.
Yield 32 bu. per acre. I observe that it stools well. Is very
early to be sown the time it was. The grain is an average. I
think it is a very good wheat.

PHILIP W. WELLS,
Long Reach, Tyler Co., W. Va.

FOURTH DISTRICT.

Variety of wheat tested, "Velvet Chaff."
Quality of seed only medium.
Sowed 1¼ bushels per acre, October 22, 1888, on heavy loam
soil, prepared by double-shovel plow. Wheat was sown and harrowed in after sown. Weather very wet at time of sowing. In
1887, the land was in corn. Yield 70 bushels per acre. In 1886
in timothy. Yield 1½ tons per acre. In 1885, in timothy. Yield
2 tons per acre. Soil is considered medium for wheat. Weather
following sowing was warm and very wet. Stand of wheat secured in the fall was fair, owing to grain of seed being small.
The wheat was bitten by frost, not to injure roots, only blades.

It was pretty dry in April, and wheat seemed to do well. Peculiarities of growth—Not as rapid growth as "German Amber." Branches more, and was thicker on ground. Ripened June 23. Cut June 20. Yield 19 bushels per acre. In favor of this. It is early and good straw. Did not rust. Grain is larger than that of seed sent and seems to be very much improved. All like it that see it. Think it a good wheat. It seem to be harder than "German Amber."

W. H. HENDERSON,
Long Reach, Tyler Co., W. Va.

FOURTH DISTRICT.

Variety of seed wheat tested, "German Amber."
Quality of seed was good.
Sowed 1¼ bushels per acre, October 2, 1888, in heavy loam soil, prepared by double-shovel plow and harrowed in. Soil at time of sowing was very wet. Land produced 70 bushels corn per acre in 1887. In 1886 timothy. Yield 1½ tons per acre. In 1885 two tons timothy per acre. Land is considered medium for wheat. Weather following sowing was warm and very wet. Stand of wheat secured in the fall was only medium. Did not branch much. Bit by frost in April, although not enough to injure it. Some rust in June. Weather pretty dry in April, and wheat seemed to do well. Pecularities of growth—Grew slow until May, then commenced growing and branching and grew very fast. Ripened first of July. Cut on the 4th. Yield 17½ bushels per acre. Think it was owing to the late sowing that the wheat did not branch until May. I think it a good wheat. The berry is large and of good quality.

W. H. HENDERSON,
Long Reach, Tyler Co., W. Va.

FOURTH DISTRICT.

Variety of wheat tested "Michigan Bronze."
Quality of seed was good.
Sowed 1½ bushels per acre October 18, 1888. Soil is a light, sandy clay, prepared by being cultivated in corn during the preceding summer. Wheat was sown on corn stubble, and plowed in with a double shovel plow. Soil at time of sowing was very wet. Land in corn in 1887. Yield 15 bushels per acre. In 1886 in oats. Yield 8 bushels per acre. In 1885 in corn. Yield 10 bushels per acre. Soil is considered very poor and unproductive. Weather following time of sowing was wet and cold. Stand of wheat secured in the fall was very inferior, caused by late sowing and bad weather. No accidents to the crop. From time of sowing to harvest weather was wet and warmer than usual.— Ripened July 3rd. Yield 9 bushels per acre. It rained almost continually during the seeding time, and it was with great diffi-

culty that I succeeded in getting the wheat sown. It was put in
in very bad condition. I consider this a very good wheat, and
was surprised that it yielded as much as it did under the very
unfavorable circumstance under which it was sown, and the difficulties it encountered.

· D. D. JOHNSON,
Long Reach, Tyler county, W. Va.

FOURTH DISTRICT.

Variety of wheat tested "Solid Gold."
Quality of seed was good.
Sowed 1¼ bushels October 30th, 1888, on chestnut oak soil prepared by being cultivated in corn the preceding summer. Wheat
was sown in corn stubble, and plowed in with a double shovel
plow. It was very wet at time of sowing; so wet that it was
almost impossible to do the seeding. The land in 1887 had nothing on part. Tobacco and potatoes on part. In 1886, uncleared.
Soil is considered poor and unproductive. Weather after sowing
was very wet and cold. The stand of wheat secured in the fall
was very inferior, caused principally by late sowing and bad
weather. The whole season was very warm and wet. The wheat
did remarkably well. The wheat grew poorly until spring, when
it made a fine growth. Ripened July 6th. Yield 10 bushels per
acre. I consider this a very good wheat, and under favorable
circumstances and in good soil, I believe it would yield 25 bushels per acre of good hard wheat.

.D. D. JOHNSON,
Long Reach, Tyler county, W. Va.

FOURTH DISTRICT.

Variety of wheat tested "Nail Club."
Quality of seed good.
Sowed 1 3-5 bushels per acre November 5, 1888, on heavy clay
soil, prepared by being cultivated in tobacco the preceding summer. The wheat was sown and plowed in with a double shovel plow.
soil at time of sowing was wet. In 1887 the land was in tobacco.
Yield 800 pounds per acre. In 1886 uncleared woods. Soil is
considered good. Weather after sowing was wet and cold.
Stand of wheat secured in the fall was not very good on account
of the lateness of the sowing. From sowing to reaping the
weather was very wet and warmer than usual. Peculiarities of
growth of crop—dark green color and "stooled" well, making a
vigorous growth in the spring. Ripened July 9th. Yield 19
bushels per acre. This wheat has a remarkably stiff straw, and
stands up well. All three of the samples sent to me were sown
on very high ground and under very unfavorable circumstances.
I consider the "Nail Club" decidedly the best and most produc-
tive wheat I ever saw. I am now sowing a 10 acre field of river

bottom land in this wheat, and expect a yield of at least 30 bushels per acre.

D. D. JOHNSON,
Long Reach, Tyler county, W. Va.

FOURTH DISTRICT.

Variety of seed tested "Solid Gold."
Quality of seed was good.
Sowed 1¼ bushels per acre October 31, 1888, on dark clay soil prepared by plowing with an ordinary plow, and harrowed in with ——— harrow. Condition of soil at time of sowing was not good, as it was too wet. In 1887 the same land was in oats. Yield 20 bushels per acre. In 1885 and 1886 in grass. Soil is considered ordinary. Stand of wheat secured in the fall was only moderate. Crop was hardy and no diseases appeared. It withstood frost and drought better than adjoining fields. Growth of crop was fair for the season. Ripened about June 28th. Yield 10 bushels per acre. In my opinion, this will be fine wheat for this country when fully developed or given a fair chance.

S. MAYFIELD,
Middlebourne, Tyler county, W. Va.

RECAPITULATION FOR THE FOURTH DISTRICT.

"GERMAN AMBER."

Considered a good wheat.

"VELVET CHAFF."

Considered a very good wheat.

"MICHIGAN BRONZE."

Considered a good wheat.

"SOLID GOLD."

Considered a good wheat.
Considered a good wheat.

"NAIL CLUB."

Considered the best wheat in the country.

19 tests were sent to this District, of which 6 were reported upon as follows:
"German Amber." 1 report favorable.
0 reports unfavorable.

173

"Velvet Chaff,"	1 report favorable.
	0 reports unfavorable.
"Michigan Bronze,"	1 report favorable.
	0 reports unfavorable.
"Solid Gold,"	2 reports favorable.
	0 reports unfavorable.
"Nail Club,"	1 report favorable.
	0 reports unfavorable.

FIFTH DISTRICT.

Variety of wheat tested, "Michigan Bronze."
Quality of seed was good.
Sowed 2 bushels per acre, November 2nd, 1888, on thin sandy soil, corn stubble. Seed was shoveled in with double shovel plow. Condition of soil at time of sowing was very wet. Soil is considered poor for wheat. Weather following sowing was very wet. Stand of wheat secured in the fall was only tolerably good.—Ripened a little earlier than the "Fultz." Yield 6 bushels per acre. Straw was light and firm. Think it a very good wheat. It is the best I ever raised.

R. L. BARR,
Spencer, Roane Co., W. Va.

FIFTH DISTRICT.

Variety of wheat tested, "Michigan Bronze."
Quality of seed was good.
Sowed 1¼ bushels per acre, November 1, 1888, on sandy loam soil in corn stubble. Soil is not considered very good for wheat. Weather following sowing was rainy. Ripened last of June. Yield 6 bushels per acre.

Same report for "Velvet Chaff."
Both kinds of wheat are better than what we have been raising. "Fultz" has been the leading wheat here.

A. E. RODABAUGH,
Spencer, Roane Co., W. Va.

FIFTH DISTRICT.

Variety of wheat tested, "German Amber."
Quality of seed was good.
Sowed 1¼ bushels on good soil, prepared by plowing and harrowing wheat in. Condition of soil at time of sowing was wet. Stand of wheat secured in the fall was not good. Think this wheat is only adapted for low or swamp land. We have tried it before. Ripened 8 to 10 days later than other wheat. Yield light. This is a good wheat for low or swamp lands and produces an

excellent quality of bread. It is not snow white, but the quality of the bread is first class.

H. DePoe,
Spencer, W. Va.

FIFTH DISTRICT.

Variety of wheat tested, "Velvet Chaff."
Quality of seed good.
Sowed 1¼ bushels per acre, October 15, 1888, on clay and sandy soil, prepared by plowing and harrowing. Condition of soil at time of sowing was wet. Corn was grown on this land in 1887. Soil is considered well adapted for wheat. Weather following sowing was very wet. Stand of what secured in the fall was very good. Crop blasted about the time of ripening. Withstand frost and drought well. Ripened about the same time as "Fultz" wheat. Yield 6 bushels per acre. I do not think this variety of wheat is good.

H. DePue,
Spencer, W. Va.

FIFTH DISTRICT.

Variety of wheat tested, "Gold"
Quality of seed was good.
Sowed 1¼ bushels per acre, October 28, 1888, on clay and sandy soil, prepared by plowing and harrowing. Condition of soil at time of sowing was wet. Land was in sod 25 years. Soil is considered good for wheat. Stand of wheat secured in the fall was only moderate. Wheat was winter killed. This wheat does not withstand frost and drought. Crop grew very tall, with long head and bright, soft, yellow straw. Ripened 10 days later than other wheat. Yield 7 bushels per acre. Think it a good wheat.

H. DePue,
Spencer, W. Va.

FIFTH DISTRICT.

Variety of seed tested, "Michigan Bronze."
Quality of seed tested was good.
Sowed 1¼ bushels per acre, October 15th, on mixed clay and sandy soil, prepared by plowing and harrowing wheat in. Soil was too wet at time of sowing on account of continual rain. Land had been in sod for 25 years. Considered well adapted for wheat. Wheat was blasted in head about the time of ripening. Ability of wheat to withstand frost and drought is good. It grows strong and vigorous, with good, strong, healthy straw. Yield was 8 bu. per acre. Think this a good wheat for this section with a fair chance.

H. DePue,
Spencer, W. Va.

175

Variety of wheat tested, "Michigan Bronze."
Sowed 1 bushel on ¾ acre of upland, loamy soil, prepared by
shoveling grain in with shovel plow. Condition of soil at time of
sowing was wet, and continued wet until harvest. Crop was damaged some in the head before harvest by wet weather. Yield 5½
bushels.

C. C. KELLY.
Spencer, W. Va.

RI CAPITUIATION FOR THE FIFTH DISTRICT.

"MICHIGAN BRONZE."

Considered good.
Considered good.
Damaged in the head.
Considered good.

" GERMAN AMBER."

Good for low lands.

" VELVET CHAFF."

Considered good.
Considered good.

" SOLID GOLD."

Considered a good wheat.

Twenty-six tests were sent to this district, of which eight were
reported upon as follows :

"Michigan Bronze," 3 reports favorable.
         1 report unfavorable.
"German Amber, 1 report favorable.
         0 reports unfavorable.
"Velvet Chaff," 2 reports favorable.
         0 reports unfavorable.
"Solid Gold," 1 report favorable.
         0 reports unfavorable.

SIXTH DISTRICT.

Variety of wheat tested, "Findlay."
Quality of seed received was good.
I sowed 1¼ bushels per acre, on October 22, 1888   Soil is a light,

sandy hill-land and was prepared by drilling in corn land with 200 pounds of Anchor Brand fertilizer per acre. Condition of soil at time of sowing was a little wet, but otherwise in good condition. Raised a good crop of corn. Land considered good for corn or wheat. Weather following was very wet. Stand of wheat secured in the fall was good. I believe it a good variety to withstand frost or drought. It is a few days later than the "Rice" or "Booten." Ripened about June 20. Yield 23 bushels per acre. I think it a good, hardy wheat and a good yielder.

R. M. SIMMS,
Scary, Putnam Co., W. Va.

SIXTH DISTRICT.

Variety of wheat tested, "Rice."
Quality of seed was good.
Sowed 1 bushel per acre, October 22, 1889, on light, sandy hill land. Prepared by sowing in corn land with drill and 200 pounds of Anchor Brand fertilizer used to the acre. Soil at time of sowing was wet and in good condition. The soil is considered good for corn or wheat. Weather following sowing was very wet. Stand of wheat secured in the fall was good. Ability of the crop to withstand frost and drought good. It ripened 3 or 4 days earlier than other varieties. Yield 4 bushels per acre. Think it a good, hardy, plump wheat, but not much of a yielder. It was sown side by side of "Findlay" and "White Booten," all on same soil.

R. M. SIMMS,
Scary, W. Va.

SIXTH DISTRICT.

Variety of wheat tested, "White Booten."
Quality of seed was good.
Sowed 1¼ bushels to the acre, October 20, 1888, on sandy, hill land, prepared by sowing broadcast and plowing in, with 200 pounds of fertilizer per acre. Condition of soil at time of sowing was wet. Had raised good potatoes upon the land. It is considered good corn or wheat land. Weather following the sowing was very wet. Stand of wheat secured in the fall was very good. Ability of wheat to withstand frost and drought is very good. Ripened June 20. Yield 14 bushels per acre. Believe to be a very good wheat.

R. M. SIMMS,
Scary, Putnam Co., W. Va.

SIXTH DISTRICT.

Variety of wheat tested "Findlay."
Quality of seed received was good.
Sowed 1¼ bushels per acre November 1, 1888, on a black clay

soil, prepared by harrowing corn stubble and drilling. Condition
of soil at time of sowing too wet. Did not have time to turn it.
Soil is considered good. Weather following was so wet, did not
secure a stand in the fall thick enough. The heads were a little
damaged. They filled at bottom, but near the point were not
well filled. Last winter was not a fair test of its ability to withstand frost and drought. It did not have time to branch out.
Ripened June 26th. Yield 8 bushels per acre. I had been sowing Lancaster on the same land some years before. Yield was
about 10 bushels per acre. The grain is harder or firmer than
either of the other varieties, but I do not know whether it is well
adapted to our country or not.

J. F. Chapman,
Hurricane, Putnam county, W. Va.

Variety of wheat tested, "White Booten."
Quality of seed was good.
Sowed 1¼ bushels on November 1, 1888. Soil was black, heavy
and prepared by light plowing and harrowing in. At time of
sowing, it was a little too wet. Soil is considered good for corn.
Weather following sowing was rainy, and the stand of wheat
secured in the fall was tolerably good. It was sown too late to
branch. We had an uncommonly mild winter; but little freezing. Time of ripening, June 25th. Yield 6 bushels per acre. I
think that upon fresh clay soil, high land, the quality of the grain
would be nice, but do not think the yield would be large.

J. F. Chapman,
Hurricane, Putnam county, W. Va.

Variety of seed wheat tested, "Rice."
Quality of seed received was good.
Sowed 1¼ bushels per acre November 1, 1888, on heavy loam
soil. It was prepared by harrowing corn stubble, and put in with
drill. The condition of the soil at time of sowing was too wet.
Stand of wheat secured in the fall was tolerably good. It was
very weedy, as the land was not fallowed. Last winter was uncommonly mild. The crop grows very tall; about five feet.
Some stalks I measured were 5½ feet. Time of ripening was
June 20th. Yield was 10 bushels per acre. I think it a very
good wheat. I could not give it a fair test last fall as I did not
get the wheat in until late and it rained almost all fall, and was
sown too late to branch. I have just threshed and could not report sooner.

J. F. Chapman,
Hurricane, Putnam county, W. Va.

SIXTH DISTRICT.

Variety of wheat tested, "Rice."
Quality of seed received was good.
Sowed 1¼ bushels per acre, November 1, 1888 on alluvial clay
soil, prepared by running a disk harrow over corn stubble and followed by drill. Soil at time of sowing was rather heavy from previous incessant rains. Land produced 50 bushels corn per acre in
1887. In 1886 and 1885 clover pasture. The soil is considered
favorable for wheat. Weather following sowing was wet and cold.
Stand of wheat secured in the fall was fair. Incessant rains while
the wheat was in shock impaired the quality. Winter was too
mild and moist to determine as to the ability of the crop to withstand frost and drought : the soil not being dry since the wheat
was sown. It grows with a strong, stiff straw, and a foot taller
than Findlay and Mediterranean which grew on either side of it.
Ripened June 24th. Yield 16 22–60 bushels per acre. I think
the wheat a good variety. Did not tiller out as well as the "Findlay ;" consequently did not yield as well.
J. K. Thompson.
Raymond City, Putnam Co., W. Va.

SIXTH DISTRICT.

Variety of wheat tested, "White Booten."
Quality of seed received was good.
Sowed 1¼ bushels per acre, November 1, 1888, on an alluvial
clay soil, prepared by running a disk harrow over corn stubble
followed by drill. Condition of soil at time of sowing was heavy
from previous incessant rains. Land produced 50 bushels of corn
in 1887. In 1886 and 1885 in clover pasture. Land is considered
favorable for wheat. Weather following was wet and mild. Stand
of wheat secured in the fall was good. It was attacked by rust
and nearly ruined. It was a complete failure. Opinion of the
wheat, no good.
J. K. Thompson,
Raymond City, Putnam Co., W. Va.

SIXTH DISTRICT.

Variety of wheat tested, "Findlay."
Quality of seed received was good.
Sowed 1¼ bushels per acre, November 1, 1888, on an alluvial
clay soil, prepared by running a disk harrow over the corn stubble followed by drill. Condition of soil at time of sowing was
rather heavy. Land produced 50 bushels corn per acre in 1887.
In 1886 and 1885 in clover. Soil is considered favorable for wheat.
Weather following sowing was wet and mild, and the stand of
wheat secured in the fall was good. No means of testing the ability of the wheat to withstand frost and drought, as the winter was

mild and wet, and the ground has not been dry since the wheat was sown. Incessant rains while in shock impaired the quality. It is a good stiff straw of medium height. Tillered well. Ripened July 1st. Yield 18 35-60 bushels per acre. I think highly of the "Findlay." It was sown and grown under unfavorable circumstances and yet made a fair yield.

J. K. THOMPSON,
Raymond City, Putnam Co., W. Va.

SIXTH DISTRICT.

Variety of wheat tested, "Findlay."
Quality of seed was fairly good.
Sowed 1¼ bushels per acre on sandy clay soil, prepared by turning and using disk harrow. Condition of soil at time of sowing was very good. Previous crop grown (1887) was clover. Soil is considered good, and the weather following was rainy. The stand of wheat secured in the fall was very good. It was attacked by green louse with considerable injury. Ability of the crop to withstand frost and drought very good. Ripens early with good head and straw. Ripened about June 15th. Yield was 17 bushels per acre. I think it a good wheat, but did not have fair test owing to the wet weather.

JAS. STEWART,
Raymond City, Putnam county, W. Va.

SIXTH DISTRICT.

Variety of wheat tested, "White Booten."
Quality of seed sent was fairly good.
Sowed 1¼ bushels per acre October 15, 1888, on sandy clay soil, prepared by turning plow and disk harrow. Condition of soil at time of sowing was very good. Land in clover in 1887. Soil is considered good. Weather following time of sowing was rainy. Stand of wheat secured in the fall was very good, but was attacked by green louse with considerable injury. The ability of the crop to withstand frost and drought is fairly good. It ripens a little later than the "Rice.' Has short head and straw.— Ripened June 10th. Yield 15 bushels per acre. Do not think this year has been a fair test, owing to the rainy season.

JAS STEWART,
Raymond City, Putnam county, W. Va.

SIXTH DISTRICT.

Variety of wheat tested, "Rice."
Quality of seed received was good.
Sowed 1¼ bushels per acre October 15, 1888, on sandy clay soil, prepared by turning plow and disk harrow. Condition of soil at time of sowing was very good. Previous crop grown (1887) was

clover. Soil is considered good for wheat. Weather following was rainy and stand of wheat secured in the fall was very good. It was attacked by a green louse. Ability of crop to withstand frost and drought is fairly good. It grows with a good straw and well headed Ripened June 15th. Yield 17 bushels per acre. This was not a fair test year owing to excess of rain. Think it a very good wheat.

JAS. STEWART,
Raymond City, Putnam county, W. Va.

SIXTH DISTRICT.

Variety of wheat tested, "Findlay."
Quality of seed was good.
Sowed at the rate of 1¼ bushels per acre, November 5, 1888, on clay soil, prepared by turning over. Condition of soil at time of sowing was splendid. It had been in timothy for 6 years previous. It is considered good wheat land. Wheat was sown entirely too late and the weather following was wet and cold. Stand of wheat secured in the fall was good for the late sowing. No accidents or diseases damaged the crop. Wheat is well adapted to withstand frost and drought. Grows tall, and I think is very good wheat for dry land. Ripened June 25th. Yield 22½ bushels per acre. Think it a good wheat.

WILL T. COX,
Cox's Landing, Cabell Co., W. Va.

SIXTH DISTRICT.

Variety of wheat tested, "Rice."
Quality of seed was fine.
Sowed at the rate of 1½ bushels per acre, November 5, 1888, on clay soil, prepared by turning over. Condition of soil at time of sowing was excellent. Land had been in timothy for 6 years. Soil is considered good wheat land. Time of sowing was entirely too late and weather following was wet and cold. Stand of wheat secured in the fall was good for time of sowing. No accident happened to the crop. Think it will stand frost well. It grows very tall, and will do well in poor, high land. Ripened June 25. Yield 21½ bushels per acre. Grains large, but stand of wheat was not so thick as "White Booten." I think it a good wheat.

WILL T. COX,
Cox's Landing, Cabell Co., W. Va.

SIXTH DISTRICT.

Variety of wheat tested, "White Booten."
Quality of seed was good.
Sowed at the rate of 1½ bushels per acre, Nov. 6, 1888, on a dark, clay soil, prepared by turning, plowing and harrowing. Wheat

sowed with a drill. Condition of soil at time of sowing very loose and in good condition. Land in timothy hay for 6 years previous. Soil is considered good wheat land. Sown entirely too late, and the weather following was wet and cold during the fall. Stand of wheat secured in the fall was good for late sowing. No accidents or diseases damaged the crop. I think it very hardy and well adapted to rich land. Has a stiff straw. Ripened June 25. Yield 22½ bushels per acre. I find that November is entirely too late to sow wheat in this country, but the seed was delayed, and wet weather still delayed until November 6.

Opinion of wheat: splendid.

WILL T. COX,
Cox's Landing, Cabell Co., W. Va.

Variety of wheat tested, "Findlay."
Quality of seed was good.
Sowed 1¼ bushels per acre, on November 1, 1888, on sandy soil, prepared by plowing wheat in with a double-shovel plow and harrowing the ground. Condition of soil at time of sowing was wet. Previous crop grown was corn; yield 40 bushels per acre. In 1886 and 1885 land was in timothy; yield 1 ton per acre. Soil is considered medium. Weather following sowing was wet. Stand of wheat secured in the fall was not as good as the other two varieties. No accidents or diseases damaged the crop. Think this variety not as good as the other varieties tested to withstand frost and drought. Ripened June 15. Yield at the rate of 9 bushels per acre. Growth not so vigorous as other varieties and not as early as other two. Early variety is the kind of wheat needed for this country. This wheat was sown on as good ground as the other two, but it never does as well sown on corn ground as on tobacco or potatoes.

R. ENSLOW,
Huntington, Cabell Co. W. Va.

Variety of seed tested, "White Booten."
Quality of seed was good.
Sowed 1½ bushels per acre, November 1, 1888, on sandy loam soil, prepared by plowing and harrowing. Land produced 100 bushels potatoes in 1887. In 1885 and 1886 it was in grass, producing one ton of hay per acre. Soil is considered mediem. Weather following sowing was dry. Stand of wheat secured in the fall was good. No accidents or diseases damaged the crop. Consider ability of wheat to withstand frost and drought good. Ripened June 10th. Yield 18 bushels per acre. I think it a good variety, but was sown later than it ought to have been. On account of wet weather did not get to sow wheat until November 1.

I think if sown last of September or first of October it would ripen about June 1.

R. Enslow.
Huntington, Cabell Co., W. Va.

SIXTH DISTRICT.

Variety of wheat tested, "Rice."
Quality of seed was good.
Sowed 1¼ bushels per acre, on November 1, 1888. Soil is a sandy loam, prepared by plowing and harrowing. Sowed broadcast and plowed in. Previous crop grown was potatoes; yield 100 bushels per acre. In 1886 in timoty; yield 1 ton per acre. In 1885 in timothy; yield 1 ton per acre. Soil is considered medium. Weather following sowing was wet, and stand secured in the fall was good. No accidents or diseases damaged the crop. Well adapted to withstand frost and drought. No peculiarities in growth of crop noted. Time of ripening was June 8. Yield per acre was 23½ bushels. Growth of the wheat was better than the "White Booten," and finest straw that I ever saw grown on wheat. I think the variety of the wheat a good one, but it was sown later than it ought to have been. Received wheat two weeks late, then it was wet for two weeks following. Did not get to sow until November 1. Think if wheat was sown last of September or first of October it would ripen June 1.

R. Enslow,
Huntington, Cabell Co., W. Va.

SIXTH DISTRICT.

Variety of wheat tested, "Rice."
Quality of seed was good.
Sowed 1¼ bushels per acre, October 31, 1888, on sandy soil, prepared by plowing and harrowing and sown by drilling. Soil was in good condition at time of sowing. Land in grass in 1887; yield light. In 1886 in wheat; yield 9 bushels per acre. In 1885 in corn; yield 40 bushels per acre. Stand of wheat secured in the fall was poor. Crop grew very tall, and lodged badly. Ripened June 20. Yield 13½ bushels per acre. Do not think it a good wheat.

W. J. Parsons.
Huntington, Cabell Co., W. Va.

SIXTH DISTRICT.

Variety of wheat tested, "Findlay."
Quality of seed was good.
Sowed 1¼ bushels per acre, October 19, 1888, on sandy soil, prepared by plowing and harrowing. Soil was in good condition at time of sowing. Land in grass in 1887. In wheat in 1886; yield

9 bushels per acre. In 1885 in corn ; yield 40 bushels per acre. Weather following sowing was cold and rainy. There were no accidents or diseases to damage crop. Withstands frost and drought well and has good straw that will not lodge. Ripened June 20. Yield 16½ bushels per acre. Think it a good wheat.

W. J. PARSONS,
Huntington, Cabell Co., W. Va.

SIXTH DISTRICT.

Variety of seed tested, "White Booten."
Sowed 1 1-4 bushels per acre, October 31. Quality of seed was good. Soil was prepared by plowing and harrowing. Wheat sown with drill. Land in 1887 was in grass. In 1886 in wheat; 9 bushels per acre. In 1885 in corn ; yield 40 bushels per acre. Stand of wheat secured in the fall was good, and stood up well. No diseases or accidents damaged the crop. Wheat was of good height and stiff straw, and did not lodge. Ripened June 20. Yield 18½ bushels per acre. I think it a good wheat, and well adapted to this section of the State.

W. J. PARSONS,
Huntington, Cabell Co., W. Va.

SIXTH DISTRICT.

Variety of seed tested, "White Booten."
Sowed 1 1-4 bushels per acre, November 5, 1888, on thin, sandy soil. Wheat was sown on corn stubble, plowed in with double shovel plow. Soil was very wet at time of sowing. Soil is not well adapted for wheat. Weather following sowing was cold and wet. Stand of wheat secured in the fall was very good, considering the late sowing. Peculiarities of growth of crop were short heads, but good grain. Yield 12½ bushels per acre.

On account of delay in receiving wheat, we were obliged to sow it on poor land, as our good ground was all in use ; hence, do not consider it a fair test. Will try this wheat again. It looks well and strong, and think it a good wheat.

DYKE BOWEN,
Cox's Landing, Cabell Co., W. Va.

SIXTH DISTRICT.

Variety of wheat tested, "Rice."
Quality of seed was good.
Sowed 1 1-4 bushels per acre, November 5, 1888, on red clay soil, prepared by shoveling and harrowing. Condition of soil at time of sowing was fair. Crop grown in 1887 was corn ; yield 30 bushels per acre. In 1886 and 1885 in clover pasture. Soil is considered fairly adapted for wheat. Weather following sowing was rainy. Stand of wheat secured in the fall was good for the late

sowing. Crop was tangled considerably on June 16. and could
not be cradled without loss. Straw seemed to be weak. Not able
to determine as to its ability to withstand frost and drought, as
the winter was very mild and wet. Did not notice any peculiarities of growth of crop. Ripened June 25, but owing to wet
weather was not cut until the 29th. Yield 11½ bushels per acre.

Have sown the 11 1-2 bushels on river bottom land that was in
corn, and along side of the "White Booten," on the 8th of October. Sowed broadcast.

Think the straw of "Rice" is too weak, but am giving it another trial. May be able to give a better and more extended
opinion next year.

GEORGE C. BOWYER,
Winfield, Putnam Co., W. Va.

SIXTH DISTRICT.

Variety of wheat tested, "Findlay."
The seed was cracked in threshing.

Sowed 1 1-4 bushels per acre, on November 5, on light clay
soil, prepared by shoveling and harrowing. Condition of soil at
time of sowing fair only. Land was in corn in 1887; yield 25
bushels per acre. In 1886 and 1885 in red clover pasture. Soil
is considered not well adapted for wheat. Weather following
sowing was rainy. Stand of wheat secured in the fall was not
very good. Was slightly tangled June 8. Cannot say as to its
ability to withstand frost and drought, as the winter was wet and
open. Noticed no peculiarities of growth of crop. Cut June 29.
Yield 6 bushels per acre. Soil was not as good as the other lots,
and seed was considerably cracked. Was thin all through. Was
not so favorably impressed with the "Findlay." as with "Booten"
and "Rice," but it had not as good chance as the others. Have
sown the 6 bushels along side of the "Rice." There was considerable amount of cockle in this seed.

GEO. C. BOWYER,
Winfield, Putnam Co., W. Va.

SIXTH DISTRICT.

Variety of Wheat tested, "White Booten."
Quality of seed was good.

Sowed 1 1-4 bushels per acre, November 5th, 1888, on red clay
soil, prepared by shoveling and harrowing with common tooth
harrow. Condition of soil at time of sowing was only fair. Crop
grown in 1887, was corn; yield about 30 bushels per acre. In
1886 and 1885 land was in red clover pasture. Soil is considered
well adapted for wheat. Weather following sowing was wet.
Stand of wheat secured in the fall was very good, considering the
late sowing. No accidents or diseases to damage the crop. Can
not say as to the ability of crop to withstand frost and drought, as

the winter was very open and wet. Noticed no peculiarities in growth of crop. Wheat ripened about June 24, but was not cut until June 29, on account of rain. Yield 15 1-2 bushels per acre. Wheat was sown too late for a fair test, although the season was favorable. Sowed the 15½ bushels October 7, 1889, on river bottom land that had been in corn. As the season was so wet, had to sow broadcast. Like the wheat from the yield of this season. If it does as well the coming season, all things being considered, will continue to sow it.

GEO. C. BOWYER,

Winfield, Putnam Co., W. Va.

RECAPITULATION FOR THE SIXTH DISTRICT.

"FINDLAY."

Good wheat.
In doubt.
Considered fair.
Considered good wheat.
Considered good wheat.
Not favorable.
Think it a good wheat.
Not considered as good as "White Booten" or "Rice."

"RICE."

Considered a good wheat, but not much of a yielder.
Think it a good wheat.
Think it a good wheat, but not so good as the "Findlay."
Think it a good wheat.
Think it a good wheat.
Do not think it a good wheat.
Think it a good wheat.
Not favorably impressed with it.

"WHITE BOOTEN."

Think it a very good wheat.
Not a good yielder.
Not favorably impressed.
Think it a splendid wheat.

SIXTH DISTRICT.

"Think it a good variety."
Think it a good wheat, and well adapted to this section.
Think it a good wheat.
Considered good.
156 tests were sent to this district, of which 24 were reported
upon as follows:

"Findlay"   5 reports favorable.
                 3 reports unfavorable.
"Rice"   6 reports favorable.
                 2 reports unfavorable.
"White Booten"   6 reports favorable.
                 2 reports unfavorable.

SEVENTH DISTRICT.

"Did not receive any wheat whatever."

JNO. H. McCREARY.
Raleigh, Raleigh Co., W. Va.

SEVENTH DISTRICT.

37 tests sent.
No reports received favorable or unfavorable.
One notice that the wheat was not received.

EIGHTH DISTRICT.

Variety of wheat tested "White Booten."
Quality of seed received was good.
Sowed about 1¼ bushels per acre November 6, 1888 on sandy
clay soil, prepared by plowing early in October. Condition of soil
at time of sowing was loose, but very wet. Land was in wheat
in 1887. Yield about 12 bushels per acre, 1886 in corn. Yield
30 bushels per acre. In 1885 in grass. Soil is considered reason-
ably well adapted to wheat. Weather after sowing was wet, and
stand of wheat secured in the fall was good. No accidents or
diseases damaged the crop. Had very little frosts and no drought.
Growth of crop: Apparently strong and healthy. Stood up well.
Ripened June 25th. Yield estimated at 16 bushels per acre. Considering time of sowing, wet condition of the ground and having
been worked too hard and fed too little, I think the wheat has done
well. From the present experiment, I think it a good variety.

D. M. RIFFE,
Alderson, Monroe Co., W. Va.

EIGHTH DISTRCT.

Variety of wheat tested "Rice."
Quality of seed was good.
Sowed 1¼ bushel per acre about Nov. 6, 1888. Soil is a mixture of sand and clay, prepared by plowing.
Condition of the soil at time of sowing was loose, but quite wet. Previous crop grown (in 1897) was wheat. Yield 35 bushels per acre. In 1886 corn, yield light. In 1885 in pasture. Soil is considered equally adapted to the various crops. Weather following sowing was favorable, and stand of wheat secured in the fall was good. No accidents or diseases damaged the crop. There were no droughts, and only lights frosts. Crop was equally flourishing the season through. Ripened about June 25. Estimated yield 15 bushels per acre.
Considering time of sowing, wet condition of the soil, and having been worked too hard and fed too little, I think the wheat has done well. From the present experiment, I think it a good variety. D. M. RIFFE,
Alderson, Monroe county, W. Va.

EIGTH DISTRICT.

Variety of seed tested "Purple Straw."
Same report as "White Booten". except not quite so tall nor such large heads. D. M. RIFFE.
Alderson, Monroe county, W. Va.

EIGHTH DISTRCIT.

Variety of seed "Purple Straw."
Quality of seed good.
Sowed 1¼ bushels per acre October 12, 1888, on yellow clay soil, prepared by drilling in corn field. Condition of soil at time of sowing very wet. Land recently in wheat. Soil is considered moderately good. Weather following was very wet, and the stand of wheat secured in the fall was very good. Crop was injured some by wet weather. Crop was not injured by frost and drought. Ripened June 25. Yield 19 bushels per acre. Wheat grew well. I like it very much. Think our soil suits it splendidly. No rust nor smut to be found. The reason I sowed 1¼ bushels per acre, I had no 5 peck dril. I used Zell's fertilizer. I wish you could recommend a better wheat fertilizer if you can. My objections to it are that it produces too much straw.
T. L. HAYNES,
Alderson, Monroe county, W. Va.

188

EIGHTH DISTRICT.

Variety of wheat tested "Rice."
Quality of seed tolerably good.
Sowed 1½ bushels per acre October 12, 1888. Soil is yellow clay
prepared by drilling in corn stubble. Condition of soil at time of
sowing was very wet. Land in 1887 was in wheat. Soil is considered only moderately good. Weather following sowing was
very wet, and stand of wheat secured in the fall was very good.
Crop was damaged some by wet weather. Not injured by drought
or frost. Ripened July 1st. Yield 15 bushels per acre. Wheat
grew well and looked fine all the time. I think it a very good
wheat, but not as good as "Purple Straw" or "White Booten."

THOMAS L. HAYNES,
Alderson, Monroe county, W. Va.

EIGHTH DISTRICT.

Variety of wheat tested "Purple Straw."
Quality of seed was good.
Sowed 1½ bushels per acre about September 25, 1888, upon black
gravelly soil, prepared by plowing, harrowing and dragging. Condition of soil at time of sowing was not very good. Too wet.
Oats was grown on the land in 1887. Yield 32 bushels per acre.
In 1886 corn. Yield 35 bushels per acre. Soil is considered
good. Weather following sowing was rainy. Stand of wheat
secured in the fall was not very good on account of oats growing in soil. Crop was injured by the fly. Can not say as to the
ability of the crop to withstand frost and drought as the winter was very open and the weather wet enough. It made a very
good growth; was of medium height. Ripened July 1st. Yield
15 bushels per acre. I think it a good variety of wheat. It
yielded well to the amount of straw, but the samples did not come
until late, and I was almost done sowing and could not give the
wheat the chance I would have liked to have given it. Our en-
tire crop was light this year; not being nearly an average, but
the wheat received from you was rather better than my own seed.

LEVI CLAYPOOL,
Fort Spring, Greenbrier county, W. Va.

EIGHTH DISTRICT.

Variety of wheat tested "White Booten."
Quality of seed received was good.
Sowed 1½ bushels per acre October 12, 1888. Soil is a yellow
clay, and was prepared by drilling in corn land. Condition of soil
at time of sowing was wet. Land had been in wheat for three
years. Soil is considered only moderately good. Weather following time of sowing was very wet. Stand of wheat secured in
the fall was very good. Crop was damaged to some extent by

wet weather. There was no drought, and it was not injured by frost. Ripened June 25. Yield 20 bushels. It is very broad blade. Died early. Stood up real well and ripened very early. I expect to sow the "White Booten" this year. I like this wheat very well. I find it proof against rust; also have not discovered one grain of smut.

THOMAS L. HAYNES,
Alderson, Monroe county, W. Va.

EIGHTH DISTRICT.

Variety of wheat tested "Rice."
Quality of seed was good.
Sowed 1½ bushels per acre on September 25. Soil is black and gravelly, prepared by plowing, harrowing and dragging. Condition of the soil at the time of sowing was not very good on account of oats growing in the soil. Previous crop (1887) was oats. Yield 32 bushels per acre. In 1886 corn. Yield 35 bushels per acre. Soil considered good. Weather following time of sowing was wet. Stand of wheat secured in the fall was not very good on account of oats growing in the soil. Crop was injured by the fly. Can not say as to its ability to withstand frost and drought as the winter was open and weather wet. This wheat grew a little taller than the "Purple Straw." Ripened July 1. Yield 12½ bushels per acre. Wheat very good quality, but did not yield quite as well as "Purple Straw."

LEVI CLAYPOOL,
Fort Spring, Greenbrier county, W. Va.

EIGHTH DISTRICT.

Variety of wheat tested "White Booten."
Quality of seed was good.
Report same as "Rice."
I did not have land enough left in the field to sow all of this variety. It is a shorter strawed wheat than the others, but yields tolerably well to the amount of straw. It is a beautiful wheat in quality, but, I think, it requires rich soil.

LEVI CLAYPOOL,
Fort Spring, Greenbrier county, W. Va.

EIGHTH DISTRICT.

Variety of wheat tested, "White Booten."
Quality of seed received was good.
Sowed 1½ bushels per acre, November 12, 1888. Soil is bottom land; sand and loam, prepared by plowing, and wheat put in with drill. Time of sowing weather was wet. Land was in corn in 1887. Yield 40 bushels per acre. In 1886 and in 1885 in meadow. Yield 1 ton per acre. Weather following sowing was

wet and warm, and stand of wheat secured in the fall was good.
Crop seriously damaged by overflow on the 19th of May. Ability
of crop to withstand frost and drought is good. Crop grew rapidly. Ripened July 1. Not threshed. Wheat was damaged so
much by overflow that I could not form anything like a difinite
conclusion in regard to it.

M. GWINN,
Green Sulphur Springs, Summers Co., W. Va.

EIGHTH DISTRICT.

Variety of wheat tested, "Rice."
Quality of seed was good.
Sowed 1½ bushels per acre Nov. 12, 1888 on bottom land, sandy
loam, prepared by plowing. Wheat put in with drill. Condition of soil at time of sowing was wet. Land was in corn
in 1887. Yield 40 bushels per acre. In 1886 and 1885 land was
in meadow. Yield 1 ton per acre. Weather following was wet
and warm. Stand of wheat secured in the fall was good. Crop
was nearly ruined by rust. It rained here 34 days between May
15 and July 1. Ability of crop to withstand frost and drought is
good. Grows rapidly. Ripened July 1. Not threshed. I think
this wheat would suit dry upland, but it is not adapted to low
land or damp localities.

M. GWINN,
Green Sulphur Springs, Summers Co., W. Va.

EIGHTH DISTRICT.

Variety of wheat tested, "Purple Straw."
Quality of seed received was good.
Sowed 1½ bushels per acre Nov. 12, 1888, on bottom land, sandy
loam, prepared by plowing. Wheat put in with drill. Land in
corn in 1887. Yield 40 bushels per acre. In 1886 and 1885 in
meadow. Yield 1 ton per acre. Weather following sowing was
wet and warm, and stand of wheat secured in the fall was very
good. Crop withstands frost and drought well. Grows rapidly.
Time of ripening July 1. Not threshed. The letter notifying me
that you had sent the wheat was delayed for some time; the
wheat, consequently, was left lying in the depot. This accounts
for the late sowing. I am satisfied that this is a good variety of
wheat, and will suit this soil and climate.

M. GWINN,
Green Sulphur Springs, Summers Co., W. Va.

EIGHTH DISTRICT.

Variety of wheat tested, "Rice."
Quality of seed was good.
Sowed 1⅓ bushels on three-fourths acre, Nov. 1, 1888, on worn

sandy soil, prepared by plowing oats stubble and drilling the
usual way. Used 200 pounds of Zells' Commercial Fertilizer.
Soil at the time of sowing was worn, and the oats badley mat-
ted on same. Oats grown on this land in 1887. Yield 20 bushels
per acre. In 1886 corn. Yield 20 bushels per acre. Soil is considered good for wheat, but worn. Weather following sowing was
wet. Stand secured in the fall was good. Crop was damaged by
the growth of oats. It withstands frost and drought well. Ripened July 1. Yield 12 bushels per acre. Owing to adverse circumstances this was not a fair test of the wheat. Will sow
same next year. Believe it to be a good wheat.

Chas. L. Davis,
Fort Spring, Greenbrier Co., W. Va.

EIGHTH DISTRICT.

Variety of wheat tested "Purple Straw."
Quality of seed was good.
Sowed 1¼ bushels per three-fourths acre on sandy, clay soil,
prepared by plow and harrow with 200 pounds of Zell's fertilizer
drilled in. Condition of soil at time of sowing was badly worn
and wet. Corn was previous crop. Yield 7 bushels per acre.
The soil was considered good for wheat before worked out. Wheat
was sown October 16, and the weather following sowing was wet.
Stand of wheat secured in the fall was scarcely perceptible. Ability of this wheat to withstand frost and drought is extra good.
Peculiarties of growth. Wonderful stooling. Ripened June 28.
Yield 20 bushels per acre. With attention, this wheat would
yield heavily, and can be recommended.

Charles L. Davis,
Fort Spring, Greenbrier county, W. Va.

EIGHTH DISTRICT.

Variety of wheat tested "White Booten."
Quality of seed was good.
Sowed 1¼ bushels per three-fourth acre October 14, 1888, on
black, loamy limestone hillside soil, prepared by shovel plow.
Sown broadcast. Condition of soil at time of sowing was wet
and worn. Corn was grown on same land in 1887. Yield 25
bushels per acre. In 1886 oats. Yield 30 bushels per acre. In
1885 corn. Yield 35 bushels per acre. Soil is considered good
for corn and grass. Weather following sowing was very wet.
Stand of wheat secured in the fall was hardly visible on account
of time and season. There were no accidents to the crop or diseases. Think it withstood frost and drought very well. Good
straw on strong land. Ripened June 24. Yield 10 bushels per
acre. Land had been run, and I do not consider this a fair test.
With proper care, I think it would be an unusually good wheat.

Chas. L. Davis,
Fort Spring, Greenbrier county, W. Va.

192

EIGH1H DISTRICT.

Variety of wheat tested "Rice."
Quality of seed was very good.
Sowed 1¼ bushels per acre on October 19, 1838 on good clay
soil, prepared by plowing, harrowing and drilled in. Condition of
soil at time of sowing was very good, but just after sowing heavy
rains drowned some of the crop. Soil is considered fair for wheat.
Weather following was wet. Stand of wheat secured in the fall
was very good considering the wet weather. Crop stood frost
well. This wheat grew about one foot higher than common wheat
and some fell down. Ripened last week in June. Yield 20 bushels per acre. Think it a splendid wheat. I sowed some of it this
fall.   MATHEW MANN,
            Fort Spring, Greenbrier county, W. Va.

EIGHTH DISTRICT.

Variety of wheat tested "Purple Straw."
Quality of seed was very good.
Sowed 1¼ bushels per acre October 19, 1888, on good clay soil,
prepared by plowing, harrowing and drilling in. Soil was in very
good condition at time of sowing. Soil is considered   fair for
wheat. Weather following sowing was very wet. Stand of wheat
secured in the fall was very good. Withstood frost well. Ripened last of June. Yield 20 bushels per acre. Think it a very
good wheat.
                                              MATHEW MANN,
            Fort Spring, Greenbrier county, W. Va.

EIGHTH DISTRICT.

Variety of wheat tested "White Booten."
Quality of seed was good.
Sowed 1¼ bushels per acre October 19, 1888, on good clay soil,
prepared by plowing, harrowing and drilling in. Condition of soil
at time of sowing was very good. Soil is considered fair for
wheat. Wheather following was wet. Stand of wheat secured in
the fall was good with the exception of a small piece. Withstood
frost very well. This wheat did not grow so tall as the other
varieties, but stood up better. Ripened last week in June.
Yield 17½ bushels. It is a very good wheat, and a prettier wheat
than the Rice.
                                              MATHEW MANN,
            Fort Spring, Greenbrier Co., W. Va.

"White Booten."
"A very good wheat.
An unusually good wheat.
In doubt.
Well pleased.
Think it a very good variety."
"Purple Straw."
"Good wheat.
Produces too much straw.
Favorable.
Considered a good variety.
Good variety.
Think it a very good wheat."
"Rice"
"Splendid wheat.
Believe it to be a good wheat.
Adapted to uplands, but not to low, damp localities.
Good quality, but yield not equal to "Purple Straw," or "White Booten."
Think it a good variety."
123 tests were sent to this district, of which 16 were reported upon as follows:

"White Booten"	4 reports favorable.
	1 report unfavorable.
"Purple Straw"	5 reports favorable.
	1 report unfavorable.
"Rice"	5 reports frvorable.
	0 unfavorable.

TENTH DISTRICT.

Variety of wheat tested "Tuscan Island."
Quality of seed was good.
Sowed 1½ bushels per acre October 15, 1888, on bottom land, corn-stubble prepared by harrowing and drilling. Condition of soil at time of sowing was wet. Corn was grown on same land in 1887, yield 60 bushels. In 1886 in corn. Soil is considered well adapted for wheat. Weather following sowing was wet. Stand of wheat secured in the fall was good. The crop blew down in patches before filling out. Stood the winter well. Ripened about July 1st, yield 21 bushels per acre. Think if some of it had not blown down, it would have made 25 bushels. I consider it a good kind of wheat.

ANDREW LUNSFORD.
Weston, W. Va.

Variety of wheat tested "Red Russian."
Variety of wheat tested "Tuscan Island."
No particulars given. Wheat not yet threshed.

E. W. Smith, Jr,
Weston, W. Va.

TENTH DISTRICT.

Variety of wheat tested, "Deitz Longberry."
Quality of seed good.
Sowed 2¼ bushels on 1¾ acres clay soil, prepared by plowing with
two-horse plow. Soil was in good condition at time of sowing.
In 1887 land was in corn and oats. In 1885 and 1886 in pasture.
Soil is considered well adapted for wheat. Stand of wheat secured in the fall was good, but short. Noticed no accidents or
diseases to damage the crop. Cannot say as to its ability to withstand frost and drought, as we had neither this year. It ripened
about July 4. Yield 13 bushels per acre. I believe Deitz Longberry is a good wheat and will sow it and the "Hybrid Mediterranean" this fall. Both of these varieties have nice, bright straw
and nice to handle.

E. W. Smith, Jr.,
Weston, W. Va.

TENTH DISTRICT.

Variety of wheat tested, "Hybrid Mediterranean."
Quality of seed good.
Sowed 2¼ bushels on 1¾ acres of clay soil, prepared by plowing
with two-horse plow. Soil was in good condition at time of sowing. Corn and oats were grown on same land in 1887. In 1886
and 1885 the land was in pasture. Soil is considered well adapted
for wheat. Weather following sowing was rainy all the time.
Stand of wheat secured in the fall was short on account of being
sown so late. No accidents or disease affected the crop. There
was no frost or drought to test its ability in this direction. No
peculiarities of growth noticed. Ripened July 4. Yield 12 bushels per acre. Think it a good wheat. Like it better than the
"Reliable." The "Hybred Mediterranean" did not yield as much
as "Deitz Longberry," but it is about as good.

E. W. Smith,
Weston, W. Va.

TENTH DISTRICT.

Variety of wheat tested, "Reliable."
Quality of seed was good.
Sowed 2¼ bushels on 1¾ acres clay soil, prepared by plowing
with two-horse plow. Soil was in good condition at time of sowing. Corn and oats grown on same land in 1887. In 1885 and
1886 in pasture. Soil is considered well adapted for wheat.

Weather following sowing was wet. Stand of wheat secured in the fall was very short on account of late sowing. Cannot say as to its ability to withstand frost and drought, as there was no drought or hard freezing. Ripened from the 1st to the 4th of July. Yield 11 bushels per acre. Think it a good wheat.

E. W. SMITH, JR.,
Weston, W. Va.

Variety of seed tested "Valley."
Quality of seed was good.
Sowed 1¼ bushels per acre October 16, 1888, on black loam soil, prepared by a shovel plow and harrowed in. Condition of soil at time of sowing was wet. Same land was in corn in 1887. Yield good. In 1886 and 18·5 in sod. Soil is considered well adapted for wheat. Weather following sowing was very wet. Stand of wheat secured in the fall was good. No accident or diseases noticed. Withstands frost and drought about as other wheat. Looks a good deal like other bearded wheat. Ripened about the first of July. Yield 4½ bushels per acre. The grain of wheat I observed was long and shriveled some. My opinion of this wheat is that it is about same as other wheat usually raised in this neighborhood.

O. A. FRETWELL,
Buckhannon, Upshur county, W. Va,

Variety of wheat tested "Reliable."
Quality of seed was good.
Sowed 1⅓ bushels per acre about October 15, 1888, on black loam soil, prepared by plowing in with shovel plow, and harrowed afterward. Condition of soil at time of sowing was wet. Part of the same land produced corn in 1887. In 1886 it was in sod. Soil is considered well adapted for wheat. Weather following sowing was wet. Noticed no accidents to crop. Withstand frost and drought about as well as other wheat. Ripened July 1st. Yield 4½ bushels per acre. Grains are plump and full. I have sown 2⅓ bushels of this wheat this fall, as I think it worthy of another test. I think the grains are some larger and nicer looking than other wheat that I have raised.

O. A. FRETWELL,
Buckhannon, Upshur county, W. Va.

Variety of seed tested "Reliable."
Quality of seed was good.
Sowed 1¼ bushels per acre October 15th., on red clay soil. Ground was prepared by plowing with a big plow. Sowed the

wheat broal cast and harrowed it in. Condition of soil at time
of sowing was wet. Land in 1887 was in wheat. Yield not very
good. In 1885 and 1886 in corn. Yield good. Soil is considered
well adapted for wheat. Weather after sowing was very wet.
Stand of wheat secured in the fall was very good. No accidents out-
except that it was a hard season on wheat. Ability to withstand
frost and drought is as good as any other wheat. Noticed no
peculiarities of growth. Ripened July 6th. Yield about ten
bushels to the acre. It is a bearded wheat, and is a heavy, large
grain. 1 intend to sow some of it this fall as a further test. Think
it a good wheat. It seems to be hardy, and looks nice.

SILAS H. BAILEY.
Buckhannon, Upshur county, W. Va.

TENTH DISTRICT.

Variety of seed tested "Valley."
Quality of seed was good.
Sowed 1⅛ bu. per acre on October 10th, on clay soil, prepared by
plowing with large plow, and put in with harrow. Condition of
land at time sowing was good. Same land produced a good crop
of corn in 1887. In 1885 and 1886 in woods. Soil is considered
well adapted for wheat. Weather following sowing was wet.
Stand of wheat secured in the fall was good. Frequent freezes
during winter killed crop, but not more so than other varieties.
Noticed nothing peculiar in growth. Ripened July 1st to 10th.
Yield 3 bushels per acre. Think it a good wheat, and have sown
my whole raising for a second test.   G. A. NEWLON,
Buckhannon, Upshur county, W. Va.

TENTH DISTRICT.

Variety of seed tested "Red Russian."
Quality of seed good.
Sowed 1⅛ bushels on dark loam soil, prepared by plowing ground
with large iron plow, then sowed the wheat broadcast and harrowed it in. Condition of soil at time of sowing was very wet.
Previous crop grown in 1887 was corn. Good crop. In 1886 and
1885 in sod. Soil is considered good. Weather following was
very wet all fall and winter. Stand of wheat secured in the fall
was good considering the lateness of sowing. Considered the crop
badly winter killed by freezing, thawing, etc. Withstands frost
and drought about as well as any other wheat. Ripened about
July 10th. Yield 4 bushels per acre. This wheat was sown in
bottom land, and the season was wet, with several hard freezing
and thawing spells of weather. Think it has been the most unfavorable season that I have ever known. Think it a very good
kind of wheat. Grains are large and nice looking.

THOS. J. FARNSWORTH,
Buckhannon, Upshur county, W. Va.

197

Variety of wheat tested "Red Russian."
Quality of seed was mixed, bearded and smooth.
Sowed 1⅓ bushels per acre on upland in 1888, on good clay soil, prepared by plow and harrow. Condition of soil at time of sowing was wet and almost muddy. Land was in corn in 1887. Yield 50 bushels per acre. In 1886 and 1885 in grass. Soil is considered good for wheat. Weather following sowing was very rainy. Stand of wheat secured in the fall was not good. Spring was very wet and wheat did not fill well. Ability of crop to withstand frost and drought is good. It grows some taller and is some later in ripening than other varieties and was quite uneven. Ripened July 4, and some later. Yield only 3 bushels per acre. It had no fair trial, as it was sown too late and was too wet. Was put in with drill, but no fertilizer was used. Do not think it as good as the "Fultz." The "Fulcaster" wheat has the reputation of being the best kind known, but have none in this country.

D. D. FARNSWORTH,
Buckhannon, Upshur county, W. Va.

Variety of seed tested "Fultz."
Quality of seed was good.
Sowed about 1½ bushels September 25, 1888, on clay soil, prepared by drilling in after corn. Condition of soil at time of sowing was wet. In 1887 the land was in corn. In 1886 and 1885 in sod. Soil is considered well adapted for wheat. Weather following time of sowing was very wet. Stand of wheat secured in the fall was tolerably good. Observed no accidents or diseases to crop. Withstands frost well. Have had no droughts. Crop grows strong and healthy. Ripened July 6. Yield 16 bushels per acre. I consider it a good quality of wheat, and hardy.

LEVI LEONARD,
Buckhannon, Upshur county, W. Va.

Variety of seed tested "Reliable."
Quality of seed received was good.
Sowed 1½ bushels per acre October 10, 1888, on red loam soil, prepared by plowing in with shovel plow. Condition of soil at time of sowing was a little wet. Same land produced corn in 1887. In 1886 wheat. In 1885 corn. Soil is considered well adapted for wheat. Weather following sowing was very wet. Stand of wheat secured in the fall was good. No accidents, diseases or other damages to crop. Withstands frost and drought well. It heads unevenly. Was a little late in ripening. Yield, 9 bushels per acre. Weather was wet and bad all winter, with some hard

freezing on open ground. Spring was very wet. I think we have had about as hard winter for wheat as I have ever seen. Wheat yielded as well as the "Fultz" wheat that I sowed in the field and same kind of soil.

P. M. TALBOT,
Ruraldale, Upshur county, W. Va.

TENTH DISTRICT.

Variety of wheat tested, "Reliable."
Quality of seed was good.
Sowed 1¼ bushels per acre, Oct. 10, 1888, on black hill-side soil, prepared by plowing in with shovel plow. Condition of soil at time of sowing was wet. In in 1887 the same land was in corn. In 1886 in corn. In 1885 in sod. Soil is considered well adapted for wheat. Weather following sowing was very wet. Stand of wheat secured in the fall was moderately good. Crop was hurt some by wet weather. Noticed no accidents or diseases to damage the crop. Withstand frost and drought as well as any other wheat. No peculiarities of growth of crop. Ripened about July 4. Yield 7 bushels per acre. We have had an open, wet, hard winter on wheat. It is a nice looking wheat, and about as hardy as any that we have in this section. The "Fultz" may be a little better.

HENRY BRAKE,
Buckhannon, Upshur Co., W. Va.

TENTH DISTRICT.

Variety of seed tested, "Fultz."
Quality of seed was good.
Sowed 1¼ bushels, Nov. 4, 1888, on clay soil, prepared by plowing in with a shovel plow. Condition of soil at time of sowing was tolerably wet. Same land was in corn in 1887. In 1886 in sod. In 1885 in sod. Soil is considered well adapted for wheat. Weather following time of sowing was very wet. Stand of wheat secured in the fall was good. No accidents or diseases damaged the crop. Ability of wheat to withstand frost and drought is about the same as other wheat grown in this section. Noticed no peculiarities of growth of crop. Ripened about July 1. Yield 11 bushels per acre. Wheat looked nice after it was threshed, and I think it is as good as any I have raised for years. Think it as well adapted to our soil and climate as any variety that we have in this section, and likely the best.

J. S. MORRISSETTE,
Lorentz, Upshur Co., W. Va.

Variety of seed tested "Red Russian."
Quality of seed was good, but mixed. One-fourth of it was bearded wheat. Sowed 1 bushel per acre about October 8, 1888, on red clay soil, prepared by sowing broadcast and plowed in with a plow (shovel). Soil at time of sowing was rather wet. In 1888 land was in corn. In 1886 and 1885 in sod. Soil is considered well adapted for wheat. Weather following sowing was very wet, and stand of wheat secured in the fall was good. Noticed no accidents or diseases to damage the crop. Stood frost well. Have had no dry weather since wheat was sowed. I noticed that the blade and stalk of this wheat were stronger than other wheat close by. Ripened about July 5. Yield 17 bushels per acre. The season has been very bad on wheat. We have had some hard freezing, and a great deal of rain. I consider it a good wheat. Grain is large and plump. I have sowed 7 bushels of it as a further test. Think it well adapted to our soil if properly put in.

G. L. CRITES
Hinklesville, Upshur county, W. Va.

RECAPITULATION FOR THE TENTH DISTRICT.

"Fultz."
"Considered well adapted to the country.
Good quality of wheat, and hardy."
"Hybrid Mediterranean."
Think it a good wheat, but it did not yield as much as "Deitz Longberry."
"Red Russian."
"Considered a good wheat.
Do not think it as good as the "Fultz."
Think it a very good wheat.
Not threshed."
"Deitz Longberry."
"Believe it to be a good wheat."
"Reliable."
"Not equal to the "Fultz."
About equal to the "Fultz."
Think it a good wheat.
Think it worth another test.
Think it a good wheat."
"Valley."
"Think it a good wheat. About the same as other wheats raised in the same neighborhood."
"Tuscan Island."
"Considered a good wheat."
64 tests were sent to this district.

15 reported upon, as follows :

"Fultz."	2 reports favorable.
	0 reports unfavorable.
"Hybrid Mediterranean."	1 report favorable.
	0 reports unfavorable.
"Red Russian."	2 reports favorable.
	2 reports unfavorable.
"Deitz Longberry."	1 report favorable.
	0 reports unfavorable.
"Reliable."	4 reports favorable.
	1 report unfavorable.
"Valley."	1 report favorable.
	0 reports unfavorable.
"Tuscan Island."	1 report favorable.
	0 reports unfavorable.

ELEVENTH DISTRICT.

But one sample of wheat was sent out, two experiments, and it was not heard from.

TWELFTH DISTRICT.

Variety of wheat tested was "Valley."
Quality of seed was good."
Sowed 1¼ bushels per acre October 25th on corn land, prepared by harrowing and shovel plow. Condition of soil at time of sowing was good. Land was new and hill land. Weather following sowing was dry. Stand of wheat secured in the fall was good. Chinch bug damaged it slightly. Crop withstands frost and drought very well. Grows just about the same as other wheat. Ripened July 8. Yield 9¼ bushels per acre. Think it a very good quality of wheat, and suits this section very well.

James Malone,
Patterson's Depot, Mineral county, W. Va.

TWELFTH DISTRICT.

November 11, 1889.
Mr. John A. Myers,

Dear Sir:—The sample of wheat received from the West Virginia Agricultural Experiment Station consisted of 2 bushels of "Red Russian." Quality of seed was good. I sowed wheat in the fall of 1888, October 1st, on 1 5-12 acres of bottom land. Land was in corn previous year. After taking the corn off, the ground was prepared by harrowing thoroughly, then drilled in with about 140 pounds of guano to the acre. Weather following sowing was fine. Wheat came up strong, with a luxuriant growth before winter set in. Crop stood the winter well until about three weeks before cutting, and was likely to yield 20 bushels to the acre, but a heavy freshet came and badly flooded it, which affected both the filling and the quality. It yielded about 13 bushels per acre. Do not think it a desirable wheat for bottom land. Ripened July 1st. "Blue Stem Mediteranean" ripened at same time. Believe it would be a good wheat for up-land. Previous crops on same land yielded about 40 bushels to the acre. . In 1886 land was in grass. Yield about 1½ tons per acre.

Hoping that this will be a satisfactory report, I am,

Yours truly,

M. T. DAVIS,
Alaska, Mineral county, W. Va.

TWELFTH DISTRICT.

Variety of wheat tested, not given.
Quality of seed was very good.
Sowed 3 pecks on one-half acre sandy creek soil, prepard by plowing and harrowing it. Soil was in a very good state of cultivation. Seed was sown Nov. 1st., and the weather following was very good for grain. Wheat was flooded in June, when in bloom. Think it a splendid winter grain. Do not consider this a fair test as it was flooded, and did not yield much. So far as I know, I think it is a very grood grade of wheat.

A. GRIMES,
Alaska, Mineral Co., W. Va.

TWELFTH DISTRICT.

Variety of wheat tested "Red Russian."
Quality of seed was very good.
Sowed 1¼ bushels on 132 rods Oct. 27th., 1888. Soil is light and loamy, and was prepared by single shovel plow on corn land. Condition of soil at time of sowing was dry, rough and broken up in large cakes. Soil is considered pretty fair for wheat. Weather following sowing was wet and freezing. Stand of wheat secured in the fall was very good, considering the late sowing. There were no accidents or diseases to damage the crop. I believe this wheat will stand frost and drought remarkably well. Crop grew strong and vigorous, and headed two weeks, before the other wheats. Ripened July 1st., 1889. Yield 16 bushels per acre. I think it will suit this section. I like this wheat very much, and think it the wheat for this part of the country.

J. M. KELLER,
Patterson's Depot, W. Va.

TWELFTH DISTRICT.

Variety of wheat tested "Tuscan Island."
Quality of seed was very good.
Sowed 1¼ bushels per acre October 20th., on heavy clay soil,

prepared by plowing harrowing and drilling in, with a dressing of barn manure. Soil at time of sowing was in good condition. Soil is considered good for wheat. Stand secured in the fall was very good. The weather was not favorable for wheat. Ripened about July 6th. Yield 15 bushels. Do not think this variety of wheat suits this part of the state.

JOHN WEBER,
Patterson's Depot, W. Va.

TWELFTH DISTRICT.

Variety of wheat tested, "Reliable."
Quality of seed was good.
Sowed 1½ bushels per acre October 15th., 1888, on slate soil, prepared by sowing on corn stubble, and put in with shovel plow. Condition of soil at time of sowing was good. Soil is considered fair for wheat. Weather following sowing was dry. Stand of wheat secured in the fall was not very good. It withstands frost and drought rather better than other varieties. Ripened about ten days earlier than others. Yield 10 bushels per acre. The only advantage that I see in planting this variety of wheat is that it is less liable to freeze out and ripens earlier than other varieties.

WILLIAM A. WAGONER,
Patterson's Depot, W. Va.

TWELFTH DISTRICT.

Variety of wheat tested. Not given.
Quality of seed was inferior in appearance.
Sowed 1 and one-half to 1 and three-fourths bushels per acre, November 1st, 1888, on warm gravelly soil, prepared by plowing, harrowing and drilling in seed. Condition of soil at time of sowing was very wet. Work was retarded by an extremely wet fall. Crop grown in 1887, not given. Yield 10 bushels per acre. Crop grown in 1886 not given. Yield 10 bushels per acre. Crop grown 1885, not given. Yield 15 bushels per acre. Soil is considered well adapted to wheat. It was sown very late in 1888. Weather following was unfavorable. Freezing and thawing alternately. Stand secured in the fall was thin and weak, on account of late sowing I suppose. There are some chinch bugs, but they were not numerous in this part of the field. Crop seemed to stand frost and drought better than other wheat. It seemed delicate and weakly at the start, but made taller and better growth, with larger and better filled heads than that along side of it. Ripened July 1st. Yield 12 to 15 bushels per acre. The soil was better adapted to wheat growing than the average of the 25 acre field in which it was grown, but not sufficiently better to account fully for the difference in yield per acre. In as much as the average on the whole field was only about 6 bushels per acre, and this part about 12 to 15 bushels per acre, I can but think it a better

wheat than the old "Lancaster," though a smaller grained and rather a brighter wheat. The straw was taller, thicker on the ground, and the heads longer and much better filled. Whole amount of product from my sowing was about 18 bushels on something over an acre.

JNO. JOHNSON,
Alaska, Mineral County, W. Va.

TWELFTH DISTRICT.

Variety of wheat tested, "Red Russian."

Sowed 1¾ bushels per acre October 1st, 1888, on gravelly clay soil, prepared by plowing twice, harrowing and drilling in 200 pounds phosphate with seed.

Previous crop grown on this land was corn.

Soil is considered well adapted for wheat.

Weather following was wet, and stand of wheat secured in the fall was good. There were no accidents or diseases to injure the crop.

Ripened about July 1st. Yield about 22 bushels per acre.

I was deceived in the appearance of this wheat, and did not thresh it in time to sow. When I did thresh it, I regretted very much that I did not see it as it was at least one-third better than the "Mediterranean" grown under same conditions.

Respectfully,
J. W. VANDIVER,
Burlington, W. Va.

TWELFTH DISTRICT.

Variety of wheat tested, "Valley." It was clean wheat.

Sowed 1½ bushels per acre October 16th, 1888, on clay loam soil, prepared by harrowing after corn was cut. Condition of soil at time of sowing was good. Same land was in pasture in 1887. In 1886 and 1885 in pasture. Weather following sowing was favorable. Stand of wheat secured in the fall was good. Lowest part of crop froze out badly. It seems to stand drought better than frost. About one third of the wheat sent was "Fultz," some "Lancaster." The rest was "Valley." Ripened about July 6th. Yield 12½ bushels. Straw was very stiff. Did not fall in wet weather. Stood straight. Believe it would be a good wheat for high land.

HENRY E. SMITH,
Patterson's Depot, W. Va.

TWELFTH DISTRICT.

PATTERSON'S CREE, W. VA., AUG. 17, 1889.

PROF. JOHN A. MYERS,

  Director W. Va. Ex. Station,

DEAR SIR: The wheat sent me last autumn for distribution in this (12th) Senatorial district was received late, after many farmers had finished seeding. There was 44 sacks containing 2¼ bushels each.

I sent 14 sacks to R. W. Gilkeson & Son, Romney, for distribution in Hampshire county. I sent 5 sacks to George E. Leps, Keyser, for distribution in the western part of this county. The remainder I disposed of as follows; the parties receiving it being all farmers, most of them living within 8 miles of this place, though some of it went to Burlington, 20 miles south. I give name and postoffice address, and will furnish results as far as I can when the grain is threshed. The wheat sown by myself was destroyed by the great flood of May 31st. My crop of 30 acres, including a sack each of Tuscan Island and Red Russian being entirely destroyed.

NAMES OF PARTIES RECEIVING GRAIN.

John Weber, Patterson's Depot, 1 sack Tuscan Island.
John Ward, Patterson's Depot, 1 sack Tuscan Island.
William Wagoner, Patterson's Depot, 1 sack Reliable.
H. E. Smith, Patterson's Depot, 1 sack Valley.
J. M. Keller, Patterson's Depot, 1 sack Red Russian.
George Short, Patterson's Depot, 1 sack Valley.
A. Grimes, Patterson's Depot, 1 sack Valley.
James Malone, Pattersen's Depot, 1 sack Valley.
J. E. Broome, Patterson's Depot, 1 sack Valley.
M. T. Davis, Alaska, 1 sack Red Russian.
Jas. H. Long, Alaska, 1 sack Reliable.
John Johnson, Alaska, 1 sack Valley.
David Vest, Alaska, 1 sack Valley.
John W. Vandiver, Burlington, 1 sack Red Russian.
J. A. Robinson, Pattersen's Depot, 1 sack Red Russian.
J. A, Robinson, Patterson's Depot. 1 sack Tuscan Island.
J. A. Robinson, Patterson's Depot, 1 sack Tuscan Island.

Eight sacks were not disposed of, and by your advice, I sold the 18 bushels contained in them, and enclose check for the proceeds as follows:

18 bushels at 95 cts	$17.20
Less freight on grain to Romney and Keyser, as per freight bills enclosed	2.10
Freight on books	.90
Check for balance enclosed	14.20
Total	$17.20

The 15 sacks of mixed seed received were too late for delivery last fall. Each sack contained three-fourths bushel, and as the most of it was light and would cover but little ground, I had some trouble in getting it taken. One sack was torn, and most of the seed lost. One sack is still on hand. I sowed one sack, which I will report later. The other sacks were distributed in March and April during my sickness, and the young men in the store did not

take the names. They remember the following, and will shorlty
find out by inquiry the names of all, if possible:
Marcus Wagoner, Alaska, 1 sack.
W. A. Wagoner, Patterson's Depot, 1 sack.
H. E. Smith, Patterson's Depot, 1 sack.
J. H. Robinson, Patterson's Depot, 1 sack.
David G. Piles, Alaska, 1 sack.
Geoige Berry, Alaska, 1 sack.

Truly, yours,

J. A. ROBINSON.

NOTE.—The freshets and other causes prevented any successful
observations upon these grasses.

J. A. MYERS,
Director

RECAPITULATION.

TWELFTH DISTRICT.

"Valley."
"Believe it to be a good wheat for high land."
"Think it a very good quality of wheat. Suits the soil."
"Red Russian."
"Believe it to be a good wheat for upland."
"Think it suits this section well."
"Consider it a good wheat."
"Tuscan Island."
"Do not think this variety of wheat suits this part of the State."
It should be remembered that the freshet destroyed the greater
portion of the crops in some sections of this district so that re-
ports could not be rendered.
"Reliable."
"Think it a good wheat."
Reports 1 Favorable.
0 Unfavorable.
Of the 44 tests sent out 11 reports were made. Of these 2 re-
ports were "Tuscan Island," from which 0 reports were favora-
ble, 2 reports were unfavorable. 3 tests of "Valley," from which
2 reports were favorable, 0 reports were unfavorable. 3 tests of
"Red Russian" from which 3 reports were favorable, 0 reports
were unfavorable. 1 favorable report for "Reliable," no unfavorable report. Two reports of "Variety" not given.

THIRTEENTH DISTRICT.

Variety of wheat tested, "Rice."
Quality of seed was poor and dirty.
Sowed 1¼ bushels per acre, on corn land, October 15, 1888.
Soil was sandy loam, prepared by harrowing after cutting corn.
Condition of soil at time of sowing was good. Corn was grown on
land in 1887. Yield 17 to 18 bushels. In 1886 corn, with same yield.
In 1885 same. In 1888 corn, yield 20 bushels. Soil is considered

well adapted for wheat. Weather following sowing was favorable and stand of wheat secured in the fall was good. The continued rains in spring caused blight. Can not tell as to its ability to withstand frost and drought, as the winter was very open. Crop ripened early. Yield 10 bushels. Think it would be better to get wheat from the North for this section. Cannot form any opinion of the wheat under the circumstances. Open winter and unusually wet spring and early summer.

ALBERT F. DAVIS,
Rippon, Jefferson Co.. W. Va.

THIRTEENTH DISTRICT.

Variety of wheat tested, "Blue Stem" (Straw).

Quality of seed was inferior; too much garlic and cockle.

Sowed 1¼ bushels per acre, October 20, 1888, on slate and limestone soil, prepared by plowing the corn up with double-shovel plow, then drilling the wheat in. Soil was in good condition at time of sowing. Land was in pasture in 1887. Weather following sowing was fair. Wheat was frozen out very much in March, as it had a northern exposure, and scabbed very much on account of wet weather. Had no drought. Noticed no peculiarities of growth. Ripened June 26. Do not know the exact yield per acre, as I was sick when wheat was harvested, and the hands did not keep it separate. I should judge it made about 15 bushels per acre. Think it a good wheat.

W. H. LEWIS,
Kabletown, Jefferson Co., W. Va.

THIRTEENTH DISTRICT.

Variety of wheat tested, "Purple Straw."

Quality of seed good."

Sowed 1¼ bushels per acre October 12, 1888, on loamy clay soil, prepared by spring tooth harrow in corn land. Condition soil at time of sowing was a little damp. Corn grown season before. Yield 50 bushels. Soil is considered well adapted for wheat. Weather following sowing was cold and not conducive to germination. Stand secured in the fall was bad. No harm until heading. Had a mild winter. Crop stood the winter and spring well. Did not make a strong growth. Ripened June 30. Yield 17 bushels per acre. Owing to wet weather in June, the wheat scabbed badly. Never saw wheat scabb- worse than this. I prefer to withhold my opinion until I try this wheat again, as all of our wheat was more or less affected by wet weather after it came in head.

W. O. NORRIS,
Kabletown, Jefferson Co., W. Va.

THIRTEETH DISTRICT.

Variety of wheat tested "White Booten."

Quality of seed indifferent.

207

Sowed 1¼ bushels October 26 on good wheat land. Weather following sowing was very wet. White wheat cannot be grown to any profit in this country.
Jno. F. Myers,
Charlestown, Jefferson Co., W. Va.

THIRTEENTH DISTRICT.

Variety of seed tested "Purple Straw."
Quality of seed was very good.
Sowed 1¼ bushels per acre October 26, 1888, on lime stone soil, prepared by sowing on corn land. Soil was in excellent condition at time of sowing. Soil is considered well adapted for wheat. Weather following sowing was very wet. Fine stand of wheat was secured in the fall. Rust and wet weather damaged the crop to some extent. Had no drought. Stood the frost very well. Ripened June 28. Yield 20 bushers per acre. I shall sow all that I raised next year. It will be better adapted to the soil, etc., than it was this year. I regard it as an excellent variety of wheat. I sowed it ten days later than I sowed the other varieties, and in less quantity per acre, and yet it ripened at same time and yielded better. My own varieties yielded only 15 bushels per acre. Would have had better yield, I think, if I had used high grade fertilizers. The wheat came late; consequently, I had to use a cheap, inferior grade of fertilizers, as I had sown my best grade.
Jno. F. Myers,
Charlestown, W. Va.

THIRTEETH DISTRICT.

Variety of seed tested "Rice."
Quality of seed was good.
Sowed 1¼ bushels per acre on limestone soil. Crop was sown on corn ground. Soil is considered good for wheat. Sown October 24, 1888. Weather following sowing was wet. Stand of wheat secured in the fall was moderately good. Rust and wet weather damaged the crop. Had neither frost nor drought. Ripened June 23. Yield about 12 bushels per acre. I think it a pretty fair variety of wheat.
John F. Myers,
Charlestown, W. Va.

THIRTEENTH DISTRICT.

Variety of wheat tested "Rice."
Quality of seed good.
Sowed 1¼ bushels per acre on October 12, on fine, loamy soil, prepared by spring tooth harrow. The land being in corn was in fine condition for seeding. Condition of soil at time of sowing was very fine. In 1888 land was in corn. In 1887 in clover. Yield 3½ bushels per acre. In 1886 and 1885 in clover. Soil is considered very fine for wheat. Weather following sowing was a

little too cool. Stand of wheat secured in the fall was very good.
Have been sowing "Rice" wheat for a number of years. Crop
was damaged by some wet weather. Too much straw. Wheat is
very hardy, will stand drought very well. Very heavy growth of
straw is the only objection to this wheat. Ripened June 25. Yield
22 bushels per acre. I have been sowing "Rice" wheat for eight
years. Have always found it very hardy. More hardy than other
varieties, but grows too much to straw in seasons like the last,
very wet. Think it a very splendid wheat where the land is not
too strong; it being always of a heavy growth, falls too much un-
less it is a dry season.

I sowed 10 bushels of "Tuscan Island" the same day on same
kind of soil. Had a fine prospect until the wet season set in. It
was very badly scabbed. Yield was 20 bushels per acre. Believe it to be a good wheat. Will try it again this fall.

C. C. CONKLYN,
Charlestown, Jefferson Co., W. Va.

THIRTEENTH DISTRICT.

Variety of seed wheat tested "Purple Straw."
Quality of seed not good. Was mixed.
Sowed 1¼ bushels per acre October 12th, 1888, on fresh loam
soil, prepared by spring tooth harrow. It being corn land, soil
was in fine condition. Corn being thoroughly cultivated. the
land was perfectly clean. Land was in clover in 1887. In 1885
in pasture. Soil is considered well adapted for wheat. Weather
following sowing was favorable, a little cool. Stand of wheat
secured in the fall was good. Fine stand all winter but ruined by
rain in summer. Stood frost well. Had no drought. Ripened
June 25th. Yield was not over 10 bushels per acre. Wheat did
well until the rain set in in May or about blooming time. It was
badly scabbed and finally killed by rust. The past season has
been no test for any wheat. All varieties were badly damaged
by rain. Do not think it a good wheat for our country. Too liable to scab and rust. Would suggest sending only one variety to
one man, but in larger quantities. Small lots are too expensive
to keep separately.

C. C. CONKLYN,
Charlestown, Jefferson Co., W. Va.

THIRTEENTH DISTRICT.

Variety of wheat tested "White Booten."
Quality of seed received was not good.
Sowed 1¼ bushels per acre October 12th on fine loamy soil, prepared by spring tooth harrow. It being corn land, soil was in fine
condition. In 1887 land was in sapling clover. In 1885 in pasture. Soil is considered very good for wheat. Weather following sowing was a little cool. Stand of wheat secured in the fall
was not quite as strong as other varieties. Crop was damaged
some by frost. Ruined by wet weather. Does not stand frost

209

well. Had no drought. Crop died off June 20th. From observations, do not think it suits our country at all. Think it a very poor wheat under any circumstances. It was killed dead with rust.

Last fall I sowed some of the old time blue stem wheat, which did very fine. Yield was 25 bushels to the acre. Very stiff straw, and believe it will be our leading wheat again.

C. C. CONKLYN,
Charlestown, Jefferson Co., W. Va.

THIRTEENTH DISTRICT.

Variety of wheat sown "White Booten."
Quality of seed was good.
Sowed 1¼ bushels per acre on limestone soil, prepaied by harrowing corn ground. Condition of soil at time of sowing was good. Crop grown in 1887 was corn. Yield 50 bushels per acre. In 1886 in hay. Yield 2 tons per acre. In 1885 in clover. Yield 20 bushels per acre. In 1884 the land was in wheat. Soil is considered well adapted for corn or wheat. Weather following sowing was good growing weather for wheat. Stand of wheat secured in the fall was very good. Nothing injured the wheat except it sprouted some in the shock by rain. It stood frost and drought very well. It is of short and slow growth. Ripened the latter part of June. Yield was 17 bushels. Wheat raised was not of as good quality as the seed sown. Grains were not as white and plump, but were of a yellow cast and shriveled. I do not think the wheat suits the limestone soil, as it does not grow as tall or as rapidly as the wheat grown here.

HENRY P. BUSEY,
Gerrardstown, Berkeley Co., W. Va.

THIRTEENTH DISTRICT.

Variety of wheat sown "Rice."
Quality of seed was good, but very filthy. Had to clean before sowing. Sowed 62 pounds on three-fourth acre of corn land. Soil is a clay limestone, prepared by harrowing well, then sowed 150 pounds of fertilizer and top dressed with manure. Condition of soil at time of sowing was good. Crop grown in 1887 was corn. Yield 60 bushels per acre. In 1886 it was in wheat. Yield 25 bushels per acre. In 1885 in clover. Soil is considered good. Stand of wheat secured in the fall was good.

Peculiarities of growth of crop: Has a very stiff straw. Head of medium length. Ripened early, about the same time as the "Fultz," which yielded about 22½ bushels. This yielded 14 bushels and 30 pounds from the 62 pounds sown. Can not say a great deal about the wheat, but will sow more this fall.

JOHN W. TABB,
Gerrardstown, Berkeley Co., W. Va.

Variety of wheat tested, " Purple Straw."
Quality of seed was good.

Sowed 1¼ bushels per acre October 15th 1888, on soapstone soil, prepared by thoroughly harrowing and manuring with 200 pounds of South Carolina phosphate. Land is of good quality. Land was in clover in 1887. Yield 1 ton. In 1886 in clover. In 1885 in wheat. Soil is considered good for wheat. Crop was sown October 15th, and weather following was very unfavorable for two weeks. Stand of wheat secured in the fall was satisfactory considering time of sowing. Can not say whether crop would withstand frost or drought, as the winter was very mild and open. Peculiarities of growth of crop: Grew vigorously. Straw was exceedingly stiff. Matured early. Ripened June 25th. Yield 20 bushels per acre or 30 pounds. While the wheat in this section lodged badly this season, the "Purple Straw" stood up well and matured a beautiful berry, and is valuable on this account. Think the "Purple Straw" well suited to any locality where "Fultz" is successfully grown. The yield under favorable circumstances would probably be heavy. Am well pleased with "Purple Straw."

HALL WILSON.
Gerardstown, Berkeley Co., W. Va.

RECAPITULATION FOR THE THIRTEENTH DISTRICT.

"Purple Straw."
"Think it a good wheat."
"Uncertain."
"Uncertain."
"Think it a good wheat."
"Consider it a good wheat."
" Rice."
' No opinion formed."
"Think it a pretty fair variety of wheat."
"Uncertain."
"Uncertain."
" White Booten."
"Wheat not satisfactory."
"Very poor wheat."
"Wheat unsatisfactory."

It should be remembered that all varieties of wheat in this district suffered from the freshet and excessive rains in the spring. From the 51 tests sent into this district, 12 reports were received.

" Purple Straw." 3 reports were favorable.
2 reports unfavorable.
" Rice." 1 reports were favorable.
3 reports unfavorable.
" White Booten." 1 reports favorable.
2 reports unfavorable.

The seed furnished to this district seems not to have been of the quality guaranteed by the dealers, and several of the parties receiving the seed complained that is was either badly mixed or badly cleaned. It was furnished by T. W. Wood & Son, of Richmond, Virginia.

II. FRUIT TREES AND SMALL FRUITS.

A circular letter similar to the one in regard to wheat was sent to each regent in regard to testing of fruits in their respective districts. Dr. Brown of the thirteenth district also desired to test garden seeds, potatoes, etc., and upon his requisition, the lists of seeds, etc., indicated were sent to him.

The lists of fruit trees and seed sent to the respective districts are included under the district to which they were sent.

I have since called for reports upon everything sent out, but with the exception of a letter from Major Bennett, which is herewith included, no report as to the condition of the trees has been received. We do not know how many of them perished during the summer, and it is yet too early for us to have any information in regard to their fruit. The lists of trees and shrubs are placed under each regent's name receiving the same, and also the list of garden seeds, potatoes, etc., sent to the thirteenth district.

We have failed to receive any reports of the tests of grass and forage plants sent out, which, no doubt, is in a manner due to the fact that the experiments were destroyed by the unprecedented freshets and heavy rains prevailing at the season of the year when they were started.

REPORT OF HON. E. A. BENNETT.

JNO. A. MYERS,
Director Agricultural Station :
DEAR SIR :

This section of West Virginia produces only a very limited variety of apples and small fruits. It may be said, as to the former that a single variety constitutes the apple crop. The "Rome Beauty," having originated in Rome Township, Lawrence county, Ohio, just across the Ohio River, opposite th mouth of the Guyandotte River, and being an excellent bearer and fair keeper, and really a very fine fruit, has monopolized the attention of growers to the almost total neglect of other varieties. The general prejudice in favor of this apple, is emphasized by the fact, that limited experiments with other fruits, suited to the season of the year to which the Rome Beauty is not suited, have, so far, failed, and quite an extensive trial made by a large fruit grower of Huntington about twenty years ago proved a failure as to a large proportion of the varieties planted. The demand for early and summer apples is urgent and the reward to the successful grower of them will be very great. It only remains to learn what varieties will succeed in this locality to stimulate their planting and growth, and assure a large inscrease in the earning capacity of the farm lands in this section of the State. It is not expected that a fall and winter variety (to which the Rome Beauty belongs), will be shown that will be much superior in many respects to, or threaten the supremacy of, that famous apple; but it is believed that such will be found as will fairly equal it and supplement it in its season, and that other varieties will be found well suited in this section and which ripen in the early and midsummer, for which particular seasons we have now scarcely any varieties at all.

With this view, after correspondence with the Director of the West Virginia Agricultural Station, and with his approval of the method proposed, I determined to enter upon such and experimental trial of various fruits as will determine just which of them are suited to this climate and locality. Being supplied by the Station with a large variety of such fruits as could be hastily selected, owing to the lateness of the season when the experiment was determined upon, and having supplemented this with other varieties since provided by myself, these have been planted on an eligible tract of upland in the foot hills near the city of Huntington. The stock supplied by the Station was received near Christmas, 1888, just preceding a snow storm that came next day after its arrival and were "heeled" in, and so remained until a rather late spring brought suitable planting weather and condition of the ground. They were then planted in the following manner: The ground, newly cleared, was broken with a two-horse barshare plow to a depth of about eight inches. Holes or places for the young trees were then dug thirty feet apart each way, to the depth of eighteen inches and about three feet square. These were partially filled with surface soil from the surrounding earth. The roots of the young trees were then immersed in a mortar, composed of a small portion of stable manure, mixed with soil and water to the consistency of "batter," until they were thoroughly covered with it, and, at once, they were set out. One person held them in an upright position and "fingered" the earth about and among the roots, while a second shoveled the earth in until the holes were filled up about the trees. Care was taken to put plenty of fine dirt closely about the small roots and to keep them well straighted out as the filling in process went on, after which the earth was well stamped in about the tree with the feet to insure firmness and thorough contact of the soil with the roots.

Very soon after planting the leaves came out well on them, and only about ten per cent. have failed to grow. Even this per cent. of loss is to be attributed to the four weeks of drought that came in April. A number of the cherry trees blossomed this year, but the blossoms aborted and brought no fruit to maturity. It is worthy of note, that a larger per cent. of these died than of any other, and, it is probable, that the better practice would have been to rub off the fruit buds to prevent blooming and its consequent exhaustion of the tree. The land was cultivated in corn.

It will necessarily require several years before any considerable number of the trees and vines thus planted will bear and mature fruit; but it is contemplated to keep the station advised of the progress of the experiment with a view to inform the general public of the most promising varieties, and, when sufficient time has elapsed to fully develop results, a detailed report ought to go out which shall serve for a guide to those requiring this class of information. If the hopes of those patronizing and forwarding the experiment are realized such information will amply repay the cost of the work and expenditure, and confirm the view entertained by them that such a concentrated method of experimentation, with the proposed reports thereof, will prove of much more value to the public than the distribution of a like number of plants in small lots where it is impossible to get reports either of progress or results.

E. A. BENNETT,
Huntington, W. Va., Oct. 1889.

LIST OF FRUIT TREES, VINES, ETC., SENT TO HON. E. A. BENNETT.

Apples.

3 Early Harvest.
3 Golden Sweet.
3 Summer Rose.
3 Chenango.
3 Gravenstein.
3 Plump Sweet.
3 Dominie.
3 Grimes Golden.
3 Monmouth Pippin.
3 Rambo.
3 R. I. Greening.
3 Tompkins King.
3 Vandevere.
3 Belleflower.
3 Early Strawberry.
3 Kes. Codlin.
3 Tetofsky.
3 Fall Pippin.
3 Jersey Sweet.

3 Canada Reinette.
3 Fallawater.
3 Mann.
3 Northern Spy.
3 Rawle's Janet.
3 Rox. Russett.
3 Twenty Ounce.
3 Wagner.
3 Tolman's Sweet.
3 Red Russian.
3 Fanny.
3 Haskel Sweet.
3 Red Beitingheimer.
3 Lady's Sweet.
3 William's Sweet.
3 Jefferies.
3 Lady.
3 Newton Pippin.

Crab Apples.

3 Paul's Imperial.
3 Coral.
3 Dartmouth.
3 Large Yellow.
3 Montreal Beauty.

3 Yellow Siberian.
3 Currant.
3 Large Red Sib.
3 Marengo.
3 Oblong.

Pears.

3 Audre Deportes.
3 Clapp's Favorite.
3 Summer Doyenne.
3 Louise Bonne.
3 Anjou.
3 Pound.
3 Bartlett.
3 Manning's Elizabeth.

3 Howell.
3 Seckel.
3 Easter Buerre.
3 Ansoult.
3 Fred. Clapp.
3 Souv d'Congress.
3 Duhamel du Monceau.
3 Becon.

Cherries.

3 Black Eagle.
3 Black Tartarian.
3 Cleveland.
3 Trancendant Black.
3 English Morello.
3 Gov. Wood.

3 Sparhawk's Honey.
3 Napoleon.
3 Early Richmond.
3 Reine Hortense.
3 Windsor.

Plums.

3 Coe's Golden Crop.
3 General Hand.
3 St. Lawrence.
3 Fellenburg.

3 Pond's Seedling.
3 De Caradem.
3 Grand Duke.

Peaches.

3 Alberge Yellow.
3 Crawford's Late.
3 Goshawk.
3 Lord Palmerston.
3 Mt. Rose.
3 Rivers E. York.
3 Schumaker.
3 Susquehanna.
3 Walburton Admirable.

3 Conkling.
3 Foster.
3 Large E. York.
3 Louise.
3 Red Cheek Mel.
3 Salway.
3 Surpasse Mel.
3 Wheatland.

Apricots.

3 Alb. de Montgamet.
3 Breda.
3 De Coulange.
3 Moorpark.

3 Red Masculine.
3 Large Early.
3 Purple.

Quinces.

3 Orange.

3 Champion.

Nectarines or Plum.

3 Boston.
3 Pitmaston Orange.

3 Lord Napier.

Grape Vines.

3 Hartford.
3 Ionia.
3 Isabella.
3 Worden.
3 Delaware.
3 Rebecca.
3 Brighton.
3 Gaertner.

3 Mrs. Princes Museat.'
3 Syrian.
3 Moore's Early.
3 Monroe.
3 Rochester.
3 Red Chasselas.
3 Museat of Alexandria.
3 Mills.

Blackberries.

6 Agawam.
6 Erie.

6 New Rochelle.
6 Wilson's Junior.

Currants.

6 Prince Albert.

6 Victoria.

Gooseberries.

3 Industry.
3 Downing.

3 Pale Red.

Raspberries.

3 Brandywine.
3 Gregg.
3 Shaffer's Colossal.
3 Cuthbert.

3 Mammoth Cluster.
3 Caroline.
3 Golden Queen.

Figs.

1 Angelique.
1 Black Provence.
1 Castle Keennedy.

1 Black Ischia.
1 Brown Ischia.

Almonds.

2 Soft shell.

Chestnuts.

2 Spanish.

2 Japan.

Mulberries.

2 New American.

Walnuts.

2 English.

2 Dwarf Prolific.

These trees were reported by Mr. Hays last June as growing and doing well.

THIRD DISTRICT—PEREGRIN HAYS, REGENT.

Name.	Residence of the Consignee.	Name of Variety.	Quantity.
Peregrin Hays, Glenville,		Apple, American Gold Pippin	10
same		" Baldwin	10
same		" Seek No Further	10
same		" Belmont	10
same		" Ben Davis	10
same		" Buckingham	10
same		" Caroline Red June	10
same		" Danveri Winter Sweet	10
same		" Dominie	10
same		" Ladies' Sweet	20
same		" Rambo	35
same		" Rome Beauty	10
same		" Vandevere	10
same		" Wine Apple	10
same		" Yellow Belleflower	10
same		" Canada Rievrelle	6
Total			191
Peregrin Hays, Glenville,		Cherries, Governor Wood	5
same		" Kentish	5
Total			10
Peregrin Hays, Glenville,		Peaches, Hale's Early	5
same		" Early York	5
same		" Newington	5
same		" Crawford's Early	10
same		" Old Mix Free	5
same		" Norris White	5
same		" Large White Cling	5
same		" Old Mix Cling	5
same		" Heath Cling	5
same		" Crawford's Late	10
same		" Late Red Rare-ripe	5
Total			65
Peregrin Hays, Glenville,		Currants, Cherry	4
same		" Goudom White	2
same		" White Dutch	2
same		" Knights Red	2
Total			10
Peregrin Hays, Glenville,		Pears, Vicar Wakefield	4
same		" Edmond	6
same		" Stevens' Genese	3
same		" Kieffer & Lecomb	10
Total			23
Peregrin Hays, Glenville,		Plums, Washington	10
same		" Damson	10
same		" Lombard	10
Total			30
Peregrin Hays, Glenville,		Orange, Quincy	5
same		Raspberries, Cuthbert	5
same		" Gregg	5
same		" Mam. Cl	5
same		" Philadelphia	5
Total			20

Peregrin Hays, Glenville,	....	Strawberries, Sharpless	12
same	....	" Triumph de Gand	12
Total	....		24
Peregrin Hays, Glenville,	....	Blackberries, Kittatenny	6
same	....	Grape Vines, Hartford	3
same	....	" Isabella	3
same	....	" Iona	3
same	....	" Worden	3
same	....	" Brighton	3
same	....	" Delaware	3
same	....	" Gauhrer	3
same	....	" Rebecca	3
same	....	" Novres Early	3
same	....	" Monroe	3
same	....	" Rochester	3
same	....	" Princess Muscat	3
same	....	" Lyrian	3
same	....	" Red Chasseless	3
same	....	" Muscat of Alexandria	3
same	....	" Mills	3
Total	....		48

Varieties of Fruit and Ornamental Trees Sent to Dr. W. W. Brown—13th District.

Apples.
6 Yellow Transparent.
3 Baldwin.
6 Delaware Winter.
6 Transcendent Crab.
Peaches.
•12 Lord Palmerston.
6 Wheatland.
6 Waterloo Globe.
3 Magnum Bonum.
3 Lemon Cling.
Pears.
12 Lawson.
6 Kieffer's Hybrid.
Cherries.
6 Schmidt's Bigereau.
Plums.
12 Kelsey's Japan.
12 Boton Japan.
12 Spaulding.
12 Prunus Simoni.
12 Prunus Pisardi.
Currents.
12 Fay's Prolific.
Gooseberries.
12 Industry.
Grapes.
12 Niagara.

12 Moore's Early.
12 Early Victor.
12 Vergennes.
Strawberries.
100 Parry.
100 Mammoth.
100 Manchester.
Blackberries.
100 Wilson's Junior.
Raspberries.
100 Cuthbert.
Quinces.
12 Champion.
6 Meeche's Prolific.
2 Arbor Vitæ Pyramidalis.
2 Silver Fir.
3 Magnolia Accuminati.
2 Irish Juniper.
2 Wier's Cut Lv'd Maple.
6 Figs.
1 Purple Beech.
6 Dwarf English Walnut.
6 Japan Chestnut.
3 Willow Diamond.
6 Japan Persimons.
3 Magnolia Conspic.
6 Russian Apricots.

III. GARDEN SEEDS.

List of Garden Seeds, etc, sent to Dr. W. W. Brown—13th District.

AMOUNT.	SEED.	VARIETY.
2 bbls.	Potatoes.	B's Superior.
20 bu.	Oats.	Welcome.
2 bu.	Corn.	Hickory King.
5 packets.	Beans.	Head Bush and L.
3 "	Beans.	Blue Podded Butter.
1 qt.	Beans.	Perfection Wax.
1 "	Beans.	Best of all Dwarf.
1 "	Beans.	King of Garden.
1 pt.	Beans.	White Creasback.
1 "	Beans.	Lazy Wife's.
1 packet.	Beans.	White Zulu.
1 "	Beet.	B's Extra Early.
2 "	Beet.	Improved Turnip.
1 "	Cabbage.	Express.
2 "	Cabbage.	Jersey Wakefield.
1 "	Cabbage.	Vandegraw.
2 "	Cabbage.	B's Surehead.
3 "	Cabbage.	Superior.
2 "	Cabbage.	Perfection Savory.
1 "	Celery.	Gold. Self Blanching.
4 qts.	Corn.	Cory.
1 "	Corn.	Amber Cream.
1 packet.	Pop Corn.	Mapledale.
2 "	Cucumber.	Peerless.
1 "	Cucumber.	Early Russian.
1 "	Cucumber.	Giant Pera.
1 "	Lettuce.	Silver Ball.
1 "	Lettuce.	Hanson.
1 "	Lettuce.	Hard Head.
1 oz.	Melon.	Emerald Gem.
1 packet.	Melon.	Perfection.
1 "	Onion.	Victoria Red.
5 "	Onion.	Red Yell. Danvers.
4 qts.	Onion.	Sets, White.
1 packet.	Onion.	Silver Skin.
3 packets.	Parsnip.	Guernsey.
1 qt.	Beans.	Orin's Improved.
1 "	Peas.	B's Extra Early.
1 "	Peas.	Quantity.
1 "	Peas.	Quality.
1 "	Peas.	American Wonder.
1 "	Peas.	Everbearing.
2 packets.	Pumpkin.	Quaker Pie.
2 "	Pumpkin.	Summer Sweet Potato.
5 "	Pumpkin.	Japanese.
3 "	Pumpkin.	St. George.

3 packets,	Radish.	Early White Turnip.
2 "	Radish.	Suprise.
3 "	Radish.	Gt. White Stuttgart.
2 "	Radish.	Earliest Turnip.
2 "	Salsify.	Sandwish Island.
2 "	Squash.	Brazil Sugar.
3 "	Squash.	Pike's Peak.
2 "	Pepper.	Ruby King.
2 "	Tomato.	Turner's Hybrid.
2 "	Tomato.	Matchless.
1 "	Tomato.	Volunteer.
2 "	Turnip.	Breadstone.
1 "	Pop Corn.	Tom Thumb.
1 "	Beans.	Earliest of All.
1 "	Pansy.	Defiance, mixed.
1 "	Phlox.	Drum. Grand, mixed.
1 "	Verbena.	Hy. Grand.
1 "	Verbena.	Musa Ensete.
1 "	Verbena.	Eutalia Japonica.
1 "		Hold's mammoth Sage.
2 "		Hydrangea Pan.
1 "		Spirea Van Houtii.
1 "	Rose.	Bon Silene.
1 "	Rose.	Etoile de Lyon.
1 "	Rose.	Mad. Margottin.
1 "	Rose.	Marie.
1 "	Rose.	Papa Gontier.
1 "	Rose.	Perles des Jardins.
1 "	Rose.	Sunset.
1 "	Rose.	Bride.
1 "	Rose.	Adam.
1 "	Rose.	Agripina.
1 "	Rose.	Annie Olivet.
1 "	Rose.	Coquette de Lyon.
1 "	Rose.	Duchess de Fl.
1 "	Rose.	Jean d'Arc.
1 "	Rose.	La France.
1 "	Rose.	Marechal Robert.
1 "	Rose.	Pierre Gollot.
1 "	Rose.	Souv. de Malmaison.

GRASSES AND FORAGE CROPS.

THIRTEENTH SENATORIAL DISTRICT.

To Dr. W. W. Brown.
1 bushel Lucerne.
1 bushel German Amber.
1 bushel Sapling,
1 bushel Red Clover.
1 bushel Alsike.

1 sack Grass Seed mixture No. 1.
1 sack Grass Seed mixture No. 2.
1 sack Grass Seed mixture No. 3.
1 sack Grass Seed mixture No. 4.
1 sack Grass Seed mixture No. 5.
1 sack Grass Seed mixture No. 6.
1 sack Grass Seed mixture No. 7.
1 sack Grass Seed mixture No. 8.
1 sack Grass Seed mixture No. 9.
1 sack Grass Seed mixture No. 10.

TWELFTH DISTRICT.

To Colonel John A. Robinson.

*See letter of Col. Robinson under twelfth district wheat.

Destroyed by Freshets.

1 bushel Lucerne.
1 bushel German Clover.
1 bushel Sapling.
1 bushel Red Clover.
1 bushel Alsike.

No. Grass Seed mixtures from 1 to 10, inclusive. 1 sack each.

TENTH DISTRICT.

To T. J. Farnsworth.

Mostly Destroyed by Freshets.

1 bushel Lucerne.
1 bushel German Clover.
1 bushel Sapling Clover.
1 bushel Red Clover.
1 bushel Alsike.

Grass Seed mixtures 1 to 10 inclusive. 1 sack each.

The Grass Seed mixtures consisted of the following:
No. 1. For light soil. For pasture.
 Red Top.
 Orchard.
 Perrennial Rye Grass.
 Red and White Clover.

No. 2. For good, medium soil. Pasture.
 Orchard.
 Perrennial Rye Grass.
 White Clover.
 Kentucky Blue Grass.
 Tall Meadow Oat.
 Red Clover.

No. 3. For strong, deep loam. Pasture.

Red Clover.
Meadow Fescue.
White Clover.
Kentucky Blue Grass.
Red Top.
Orchard.

No. 4. For moist bottom land.   For pasture.
Italian Rye Grass.
Red Top.
Meadow Fescue.
Timothy.
Alsike.
White Clover.

No. 5. For wet bottom land.   Pasture.
Red Top.
Meadow Fescue.
White Clover.
Italian Rye Grass.

No. 6. For light soils.   Meadows.
Orchard.
Perrennial Rye.
Tall Meadow Oats Grass.
Red Clover.

No. 7. For good medium soils.   Meadow.
Tall Meadow Oat.
Kentucky Blue Grass.
Perrennial Rye.
Orchard Grass.
Red Clover.

No. 8. For strong deep loam.   Meadow.
Kentucky Blue.
Tall Meadow Oat Grass.
Red Clover.

No. 9. For moist bottom lands.   Meadow.
Red Top.
Meadow Fescue.
Timothy.
Alsike.

No. 10. For wet bottom lands.   Subject to overflow.   Meadow.
Red Top.
Meadow Fescue.

MISCELLANEOUS.

Variety of oats sown " Hargett's White."

Sowed 2 bushels per acre on a light, clay soil, fertilized at the rate of 150 pounds per acre with South Carolina bone.   Oats was sown on a plat of land in a field sown to our common oats and received the same cultivation. Crop made a vigorous growth. Ripened July 20th, ten days earlier than our common oats. Yield at the rate of 40 bushels per acres, while common oats yielded only 20 bushels per acre. Seed sown was of good quality, and I think the "Harfett Winter Oats is adapted to our soil and climate, and I can recommend this variety to our farmers. Will sow this variety another season.

Variety of oats sown, "Early Piasa Queen."

Was planted May 18, 1889, on a light sandy, loam soil, fertilized at the rate of 200 pounds per acre, with fertilizer received at the West Virginia Experiment Station. Saed did not germinate well. Plants made a sickly growth and did not mature until late in September. I would not recommend this variety on a sandy soil.

Variety of oats sown, "Early Orange."

Sown May 18, 1889, on clay soil, and fertilized with South Carolina bone at the rate of 150 pounds per acre.

Crop made a vigorous growth and ripened about September 1, 1889. Yield 80 gallons per acre. Ground was ploughed, harrowed, crossed two feet, and was cultivated three times during the growing season. I can recommend the Early Orange as suited to our soil and climate.

L. SHOMAKER.

Dellslow, Monongalia County, W. Va.

Corn.

Variety tested, "White Pearl."

Time of sowing, May 17th, on black, loose loam."

It did not mature right. Corn was soft, but I will not condemn it until I hear from others.

Oats.

Variety tested, "American."

Sowed May 8th. Crop grew finely; looked well, but unfortunately I sowed crop too near the house and it was destroyed by chickens until it was not worth saving. Crop was sown on black loam.

Planted some potatoes on ground prepared by using phosphate, but I do not think they were any better than some I planted without. I sowed the rest of the phosphate on wheat, and can not say now what the outcome will be. I have experimented a little with stable manure. The best results that I have is by harrowing manure in with either wheat or oats.

I did not plant the cane seed. I gave about half of it to my neighbors, and hope that their experience has been better than mine.

Respectfully, yours,

C. HOLLAND,

Dellslow, Monongalia County, W. Va.

The phosphate received by me did not appear to be adapted to the soil to which I applied it. I applied it on oats in alternate strips with the South Carolina bone, with bad results. The bone on my land is far superior. I also tried it on corn, also upon potatoes with the same results; the South Carolina bone being superior on all crops. The land was clay soil, with tight sub-soil. I can not recommend the phosphate.

Yours respectfully,

L. SHOMAKER,

Dellslow, Monongalia Co.

CONCLUSION.

In conclusion I may say that we have received reports from only about 15½ per cent. of the samples of wheat sent out. We have one report from experiments with small fruits, berries, etc.

The work of the Station was done with as much care as was possible under the circumstances, and we are quite positive that most of the seed wheat, fruit trees and garden seed reached their destination, as freight bills have been received by us from, and paid for, the parties receiving the goods. Almost all of the farmers who received these articles are personally unknown to the Director. They were selected by the members of the Board of Regents as being among the farmers in their districts most likely to carry out the work contemplated.

Much has been said in the agricultural papers of the country of the impractical character of the experiments conducted at the Stations over the country, and it was hoped that we would have some results in this costly experiment which would justify the claims so loudly set forth, that the work done at Experiment Stations should be entrusted to those popularly known as practical farmers. At an expense of about $2,600, the experiments so loudly called for by writers in agricultural papers have been carried out in this State with results that are anything but encouraging, the farmers failing to respond.

In looking over the reports, we find but one from 708 experiments in wheat in which any attempt was made to carry out instructions. We have failed to receive any report from 85 per cent of the experiments and very few of those making reports claim to have any more than approximations. In the most of cases, the land and crops were estimated, not measured. It generally happens that there is no unanimity in results reported upon the same varieties, so that our conclusion is that no reliance whatever can be placed upon experimental work entrusted to persons unfamiliar with scientific methods and not provided with facilities for measuring their crops and lands.

It is evident that the attention to details necessary to make an experiment successful can not be bestowed upon it by the average farmer whose time is taken up with questions of more immediate and vital interest to him, and it is hoped that our experience will be of the greatest possible value in this direction not only to this State, but to the stations located in other States. Few boards of control or boards of regents would have had the nerve to attack the problem upon the scale that our board has, and its experience if profited by throughout the country, is certainly well worth the money that has been expended in the effort. The attempt has been honestly made and faithfully carried out. It was intended as an answer to what appeared to be a popular demand that experimental work should be widely distributed over the State, and should be entrusted to "practical farmers" rather than to scientific men carefully trained in the methods of experimentation. Our board, after careful consideration of the subject and counsel with the Department of Agriculture, has deemed it wise to discontinue the effort to carry on experimental work of the character reported in this bulletin, and it has determed to abandon it.

Parties wishing seed wheat, fruit trees, etc., should therefore not expect to secure them from the station, as we have none for distribution, and they are respectfully referred to the dealers in these articles.

JOHN A. MYERS,

Director.

www.ingramcontent.com/pod-product-compliance
Lightning Source LLC
Chambersburg PA
CBHW022001190326
41519CB00010B/1352